ESTATE PU

CW00742156

BERKSHIRE

**Street Maps of 36 Towns
with index to streets**

**Road Map with index
Population Gazetteer
Administrative Districts**

Street plans prepared and published by ESTATE PUBLICATIONS and based upon the ORDNANCE SURVEY maps with the sanction of the controller of H.M. Stationery Office. Crown copyright reserved.

The publishers acknowledge the co-operation of the local authorities of towns represented in this atlas.

© Estate Publications, Bridewell House, Tenterden, Kent.

186C

ISBN 0 86084 311 4

ESTATE PUBLICATIONS

STREET ATLASES

ASHFORD, TENTERDEN
BASILDON, BRENTWOOD
BASINGSTOKE, ANDOVER
BOURNEMOUTH, POOLE, CHRISTCHURCH
BRIGHTON, LEWES, NEWHAVEN, SEAFORD
BROMLEY (London Borough),
CHELMSFORD, BRAINTREE, MALDON, WITHAM
CHICHESTER, BOGNOR REGIS
COLCHESTER, CLACTON
CRAWLEY & MID SUSSEX
DERBY, HEANOR, CASTLE DONNINGTON
EDINBURGH
EXETER, EXMOUTH
FAREHAM, GOSPORT
FOLKESTONE, DOVER, DEAL
GLOUCESTER, CHELTENHAM
GRAVESEND, DARTFORD
GUILDFORD, WOKING
HASTINGS, EASTBOURNE, HAILSHAM
HIGH WYCOMBE
I. OF WIGHT TOWNS
LEICESTER
MAIDSTONE
MANSFIELD
MEDWAY, GILLINGHAM
NEW FOREST
NOTTINGHAM, EASTWOOD, HUCKNALL, ILKESTON
OXFORD
PLYMOUTH, IVYBRIDGE, SALTASH, TORPOINT
PORTSMOUTH, HAVANT
READING
REIGATE, BANSTEAD, REDHILL
RYE & ROMNEY MARSH
ST. ALBANS, WELWYN, HATFIELD
SALISBURY, AMESBURY, WILTON
SEVENOAKS
SHREWSBURY
SLOUGH, MAIDENHEAD
SOUTHAMPTON, EASTLEIGH
SOUTHEND-ON-SEA
SWALE (Sittingbourne, Faversham, I. of Sheppey)
SWINDON
TELFORD
THANET, CANTERBURY, HERNE BAY, WHITSTABLE
TORBAY
TUNBRIDGE WELLS, TONBRIDGE, CROWBOROUGH
WATFORD, HEMEL HEMPSTEAD
WINCHESTER, NEW ALRESFORD
WORTHING, LITTLEHAMPTON, ARUNDEL

COUNTY ATLASES

AVON
AVON & SOMERSET
BERKSHIRE
CHESHIRE
CORNWALL
DEVON
DORSET
ESSEX
HAMPSHIRE
HERTFORDSHIRE
KENT (64pp)
KENT (128pp)
OXFORDSHIRE
SHROPSHIRE
SOMERSET
SURREY
SUSSEX (64pp)
SUSSEX (128pp)
WILTSHIRE

LEISURE MAPS

SOUTH EAST (1:200,000)
KENT & EAST SUSSEX (1:150,000)
SURREY & SUSSEX (1:150,000)
SOUTHERN ENGLAND (1:200,000)
ISLE OF WIGHT (1:50,000)
WESSEX (1:200,000)
DEVON & CORNWALL (1:200,000)
CORNWALL (1:180,000)
DEVON (1:200,000)
DARTMOOR & SOUTH DEVON COAST (1:100,000)
GREATER LONDON (1:80,000)
EAST ANGLIA (1:250,000)
THAMES & CHILTERNS (1:200,000)
COTSWOLDS & WYEDEAN (1:200,000)
HEART OF ENGLAND (1:250,000)
WALES (1:250,000)
THE SHIRES OF MIDDLE ENGLAND (1:250,000)
SHROPSHIRE, STAFFORDSHIRE (1:200,000)
SNOWDONIA (1:125,000)
YORKSHIRE & HUMBERSIDE (1:250,000)
YORKSHIRE DALES (1:125,000)
NORTH YORK MOORS (1:125,000)
NORTH WEST ENGLAND (1:200,000)
ISLE OF MAN (1:60,000)
NORTH PENNINES & LAKES (1:200,000)
LAKE DISTRICT (1:75,000)
BORDERS OF ENGLAND & SCOTLAND (1:200,000)
BURNS COUNTRY (1:200,000)
ISLE OF ARRAN (1:63,360)
ARGYLL & THE ISLES (1:200,000)
HEART OF SCOTLAND (1:200,000)
GREATER GLASGOW (1:150,000)
LOCH LOMOND & TROSSACHS (1:150,000)
PERTHSHIRE (1:150,000)
FORT WILLIAM, BEN NEVIS, GLEN COE (1:185,000)
IONA (1:10,000) & MULL (1:115,000)
GRAMPIAN HIGHLANDS (1:185,000)
LOCH NESS & INVERNESS (1:150,000)
AVIEMORE & SPEY VALLEY (1:150,000)
SKYE & LOCHALSH (1:130,000)
CAITHNESS & SUTHERLAND (1:185,000)
WESTERN ISLES (1:125,000)
ORKNEY & SHETLAND (1:128,000)
ENGLAND & WALES (1:650,000)
SCOTLAND (1:500,000)
GREAT BRITAIN (1:1,100,000)

ROAD ATLAS

MOTORING IN THE SOUTH (1:200,000)

EUROPEAN LEISURE MAPS

EUROPE (1:3,100,000)
BENELUX (1:600,000)
FRANCE (1:1,000,000)
GERMANY (1:1,000,000)
GREECE & THE AEGEAN (1:1,000,000)
IRELAND (1:625,000)
ITALY (1:1,000,000)
MEDITERRANEAN CRUISING (1:5,000,000)
SCANDINAVEA (1:2,600,000)
SPAIN & PORTUGAL (1:1,000,000)
THE ALPS (1:1,000,000)
THE WORLD (1:35,000,000)
THE WORLD FLAT SHEET

ESTATE PUBLICATIONS are also
sole distributors in the U.K. for:
ORDNANCE SURVEY, Republic of Ireland
ORDNANCE SURVEY, Northern Ireland

Catalogue and prices from ESTATE PUBLICATIONS,
Bridewell House, Tenterden, Kent TN30 6JB.
Tel: 05806 4225   Fax: 05806 3720

CONTENTS

BERKSHIRE ROAD MAP:
(Approx. 4½ miles to 1 inch)

pages 4—5

GAZETTEER INDEX TO ROAD MAP:
(with populations)

pages 6—7

BERKSHIRE ADMINISTRATIVE DISTRICTS:

pages 8—9

TOWN CENTRE STREET MAPS:

Arborfield	page 10	Old Windsor	page 45
Arborfield Cross	10	Pangbourne	30
Arborfield Garrison	10	Purley/Tilehurst	31
Ascot	12	Reading	32-33
Bracknell	14-16	Sandhurst	34-35
Bray	17	Shinfield	42
Burghfield Common	17	Slough	36-37
Cookham	19	Sunningdale	13
Cookham Dean	18	Sunninghill	12
Cookham Rise	19	Thatcham	38-39
Crowthorne	20-21	Twyford	40
Datchet	22	Uxbridge (Gtr. London)	41
Eton	47	Wargrave	42
Hungerford	23	Whitchurch	30
Lambourn	22	Wick Hill	43
Maidenhead	24-25	Winnersh	44
Newbury	26-29	Windsor	46
North Ascot	11	Wokingham	48-49

INDEX TO STREETS

pages 50-64

One-way street	→	Post Office	●
Pedestrian Precinct	▨	Public Convenience	ⓒ
Car Park	Ⓟ	Place of Worship	+

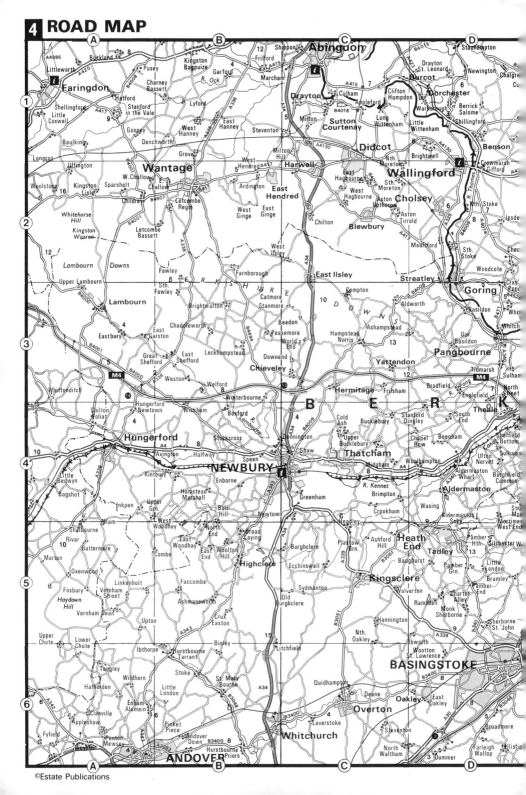

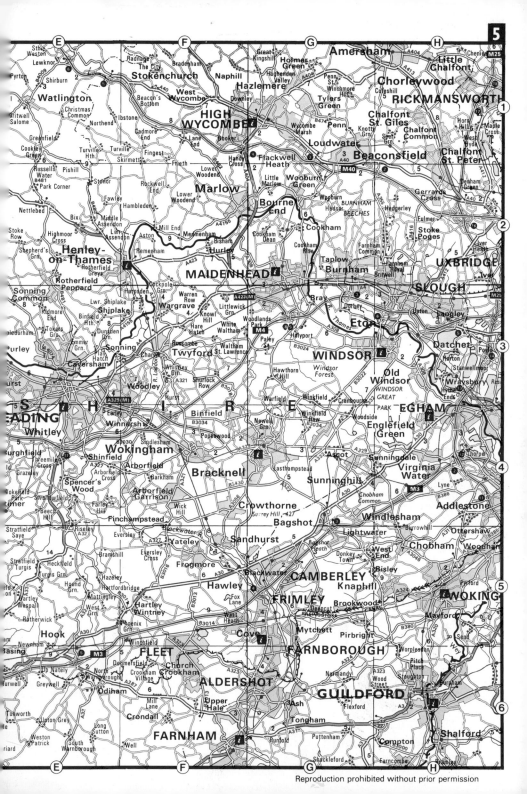

GAZETTEER INDEX TO ROAD MAP
With populations

County of Berkshire population **707,023**

Districts:
Bracknell **82,512**
Newbury **121,398**
Reading **133,540**
Slough **97,389**
Windsor & Maidenhead **131,284**
Wokingham **140,900**

Aldermaston **2,177**	4 D4	Cookham Rise	5 G2
Aldermaston Soke	4 D4	Cranbourne	5 G3
Aldermaston Wharf	4 D4	Crookham	4 C4
Aldworth **196**	4 C3	Crowthorne E.C. Wed **8,014**	5 F4
Arborfield (incl. Arborfield		Datchet E.C. Wed **3,737**	5 H3
Cross) **2,323**	5 F4	Donnington (with Shaw)	
Arborfield Cross	5 E4	**1,722**	4 C4
Arborfield Garrison	5 F4	Downend	4 C3
Ascot (Sunninghill)		Earley **11,957**	5 E4
E. C. Wed **10,305**	5 G4	Eastbury	4 A3
Ashampstead **369**	4 C3	East Garston **542**	4 A3
Aston	5 F2	Easthampstead **1,222**	5 G4
Avington	4 B4	East Ilsley E.C. Wed **394**	4 C2
Barkham **2,841**	5 F4	East Shefford	4 B3
Basildon **1,356**	4 D3	Enborne **535**	4 B4
Beech Hill **313**	5 E4	Englefield **169**	4 D3
Beedon **361**	4 C3	Eton E.C. Sat **3,559**	5 G3
Beenham **939**	4 D4	Farley Hill	5 E4
Binfield E.C. Wed **3,335**	5 F4	Farnborough **96**	4 B2
Bisham **1,079**	5 F2	Fawley **122**	4 B2
Boxford **438**	4 B4	Finchampstead **5,697**	5 F4
Bracknell E.C. Wed **49,024**	5 F4	Frilsham **327**	4 C3
Bradfield **1,416**	4 D3	Grazeley **11**	5 E4
Bray E.C. Wed **9,427**	5 G3	Gt. Shefford **631**	4 B3
Brightwalton **258**	4 B3	Greenham **2,642**	4 C4
Brimpton **512**	4 C4	Halfway	4 B4
Britwell **5,462**	5 H2	Hampstead Norris **593**	4 C3
Bucklebury **2,322**	4 C4	Hampstead Marshall **221**	4 B4
Burghfield **3,942**	5 E4	Hare Hatch	5 F3
Burghfield Common	4 D4	Hawthorn Hill	5 G3
Burghfield Hill	4 D4	Hermitage E.C. Wed **955**	4 C3
Calcot	4 D3	Holyport	5 G3
Catmore **26**	4 B3	Horton **851**	5 H3
Caversham	5 E3	Hungerford E.C. Thurs **4,855**	4 A4
Chaddleworth **513**	4 B3	Hungerford Newtown	4 A4
Chapel Row	4 D4	Hurley E.C. Thurs **2,068**	5 F2
Charvil **1,875**	5 F3	Hurst (St. Nicholas) **1,254**	5 F3
Chieveley **2,030**	4 C3	Inkpen **701**	4 A4
Cockpole Grn	5 F3	Kintbury E.C. Thurs **2,655**	4 B4
Cold Ash **1,855**	4 C4	Knowl Hill	5 F3
Combe **38**	4 B5	Lambourn E.C. Wed **3,520**	4 A3
Compton **1,247**	4 C3	Langley	5 H3
Cookham E.C. Wed **5,998**	5 G2	Leckhampstead **327**	4 B3
Cookham Dean	5 G2	Littlewick Grn	5 F3

6

Maidenhead E.C. Thurs
49,350 — 5 G2
Midgham **327** — 4 C4
Mortimer — 4 D4
Newbury E.C. Wed **26,573** — 4 B4
Newell Grn — 5 G4
North Street — 4 D3
Old Windsor E.C. Wed **5,263** — 5 H3
Padworth **395** — *
Paley St — 5 G3
Pangbourne E.C. Thurs **2,664** — 4 D3
Peasmore **201** — 4 B3
Popeswood — 5 F4
Purley E.C. Wed **4,081** — 5 E3
Reading E.C. Wed **127,736** — 5 E4
Remenham **514** — 5 F2
Risely — 5 E4
Ruscombe **1,071** — 5 F3
Sandhurst E.C. Wed **11,264** — 5 F5
Shaw cum Donnington **1,722** — 4 C4
Sheffield Bottom — 4 D4
Shinfield **6,435** — 5 E4
Shottesbrooke **144** — *
Shurlock Row — 5 F3
Sindlesham — 5 F4
Slough E.C. Wed **97,389** — 5 H3
Sonning E.C. Wed **1,469** — 5 E3
South End — 4 D4
South Fawley — 4 B3
Speen E.C. Thurs **2,456** — 4 B4
Spencer's Wood — 5 E4
Stanford Dingley **226** — 4 D4
Stanmore — 4 C3
Stockcross — 4 B4
Stratfield Mortimer **3,483** — 4 D4
Streatley E.C. Wed **1,055** — 4 D2
Sulham **82** — 4 D3
Sulhamstead **1,745** — 4 D4
Sulhamstead Bannister **88** — *
Sunningdale E.C. Wed **2,893** — 5 H4
Sunninghill E.C. Wed **10,305** — 5 G4
Swallowfield **2,236** — 5 E4

Thatcham E.C. Wed **14,847** — 4 C4
Theale E.C. Wed **6,657** — 4 D4
Threemile Cross — 5 E4
Tidmarsh **290** — 4 D3
Tilehurst **12,575** — 4 D3
Twyford E.C. Wed **4,997** — 5 F3
Upper Basildon — 4 D3
Upper Bucklebury — 4 C4
Upper Green — 4 B4
Upper Lambourn — 4 A3
Upton — 5 H3
Upton Nervet **212** — 4 D4
Waltham St. Lawrence **1,385** — 5 F3
Warfield **1,613** — 5 G3
Wargrave E.C. Wed **4,081** — 5 F3
Warren Row — 5 F3
Wasing **55** — 4 D4
Welford **540** — 4 B3
West Ilsley **331** — 4 B2
Weston — 4 B3
West Woodhay **99** — 4 B4
Whistley Green — 5 F3
Whitchurch — 4 D3
White Waltham
E.C. Thurs **3,004** — 5 F3
Whitley — 5 E4
Wickham — 4 B4
Windsor E.C. Wed **28,826** — 5 G3
Winkfield E.C. Wed **8,040** — 5 G3
Winkfield Row — 5 G4
Winnersh **6,215** — 5 F4
Winterbourne **145** — 4 B3
Wokefield Park **121** — 5 E4
Wokingham E.C. Wed **24,558** — 5 F4
Woodlands Park — 5 G3
Woodley **27,493** — 5 F3
Woodside — 5 G4
Woolhampton **568** — 4 D4
World's End — 4 C3
Wraysbury E.C. Wed **3,472** — 5 H3
Yattendon E.C. Wed **241** — 4 C3

Population figures (in bold type) are based upon the 1981 census and relate to the local authority area or parish as constituted at that date.
Places with no population figure form part of a larger local authority area or parish. District boundaries are shown on pages 8-9.
E.C. — Early Closing (not all shops close on these days).

* Parish not shown on map pages 4-5 due to limitation of scale.

Shippon
Abingdon
Stadhampton

Buckland
Frilford
Kingston Bagpuize
Garford
Drayton St. Leonard
Newington
Chalgro
Cu.

Littleworth
Pusey
Marcham
Burcot

Faringdon
Charney Bassett
Drayton
Culham
Clifton Hampden
Dorchester

Hatford
Appleford
Warborough
Berrick Salome

Shellingford
Stanford in the Vale
Lyford
Milton
Sutton Courtenay
Long Wittenham
Little Wittenham
Shillingford

Little Coxwell

Goosey
East Hanney
Steventon

Baulking
Denchworth
West Hanney

Longcot
Grove
Milton Hill
Didcot
Benson

Uffington
Wantage
West Hendred
Harwell
Nth. Moreton
Brightwell
Crowmarsh Gifford

Woolstone
Kingston Lisle
W.Challow
Sparsholt
E. Challow
Ardington
East Hendred
East Hagbourne
Wallingford

Childrey
Letcombe Regis
West Hagbourne
Sth. Moreton
Aston Upthorpe
Cholsey

West Ginge
East Ginge
Aston Tirrold
Nth. Stoke

Letcombe Bassett
Chilton
Blewbury
Ipsder

West Isley
Moulsford
Sth. Stoke
Check

Fawley
Farnborough
East Ilsley
Streatley
Woodcote

Upper Lambourn
Sth. Fawley
Compton
Goring
Cray's Pond

Lambourn
Catmore
Stanmore
Aldworth
Basildon
White H.

Brightwalton
Whitch

Eastbury
East Garston
Chaddleworth
Beedon
Peasemore
World's End
Hampstead Norris
Ashampstead
Upr. Basildon
Pangbourne

Great Shefford
East Shefford
Leckhampstead
Downend
Chieveley
Yattendon
Tidmarsh
Sulham

Weston
Welford
Hermitage
Frilsham
Bradfield
Englefield
North Street

Whittonditch
Winterbourne
NEWBURY
Theale

Chilton Foliat
Hungerford Newtown
Wickham
Boxford
Cold Ash
Bucklebury
Stanford Dingley
South End
Sheffield Bottom

Hungerford
Stockcross
Donnington
Upper Bucklebury
Chapel Row
Beenham
Ufton Nervet
Sulhams

Froxfield
Avington
Halfway
Shaw
Midgham
Thatcham
Woolhampton
Aldermaston Wharf
Burghfield Common

Little Bedwyn
Kintbury
Speen
NEWBURY
Enborne
Aldermaston

Bagshot
Hamstead Marshall
Greenham
Brimpton
Wasing
Stra'
Mor'

kpen
Upper Grn.
Ball Hill
Newtown
Crookham
Aldermaston Soke
Mortimer West End

Ham
West Woodhay
North End
Headley
Pamber Hth.
Silchester We

Shalbourne
Rivar
East Woodhay
Broad Laying
Plastow Grn.
Ashford Hill
Heath End
Tadley

Buttermere
Combe
East Woolton End
Hill
Burghclere
Baughurst
Pamber Grn.
Little London

Marten
Highclere
Ecchinswell
Kingsclere
Bramley

Oxenwood
Linkenholt
Faccombe
Sydmonton
Wolverton
Charter Alley
Pamber End

Fosbury
Vernham Street
Ashmansworth
Old Burghclere
Ramsdell
Monk Sherborne

Vernham Dean
Upton
Crux Easton
Hannington
Sherborne St. John

Upper Chute
Lower Chute
Binley
Litchfield
Nth. Oakley
Ibworth
Wootton St. Lawrence

Ibthorpe
Hurstbourne Tarrant
BASINGSTOKE

Tangley
Stoke
St. Mary Bourne
Quidhampton

Hatherden
Wildhern
Little London
Deane
Oakley
East Oakley

Clanville
Enham-Alamein
Overton

Appleshaw
Picket Piece
Laverstoke
Broadmere

Fyfield
Penton Mewsey
Andover Down
Whitchurch
Steventon
Farleigh Wallop
Ellisfiel

Weyhill
Hurstbourne Priors
North Waltham
Dummer

ANDOVER

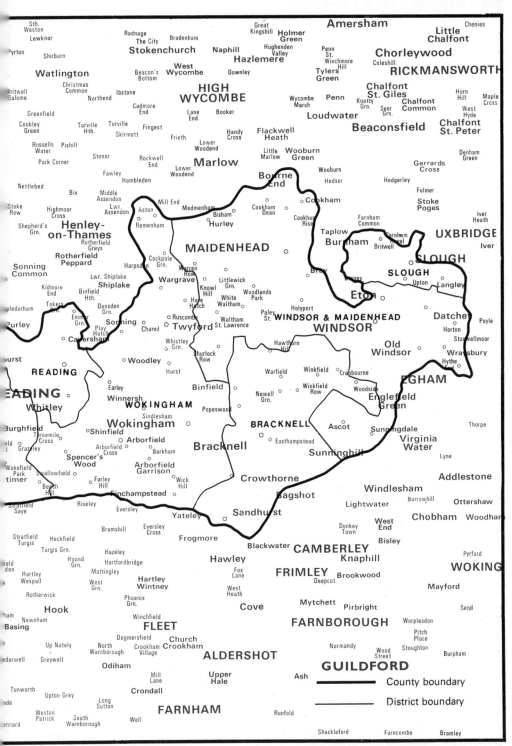

Sth. Weston
Lewknor
Pyrton
Shirburn
Watlington
Britwell Salome
Greenfield
Cookley Green
Russells Water
Park Corner
Nettlebed
Stoke Row
Highmoor Cross
Shepherd's Grn.
Henley-on-Thames
Sonning Common
Kidmore End
Binfield Hth.
Tokers Grn.
pledurham
Emmer Grn
Purley
Caversham

Radnage
The City
Stokenchurch
Christmas Common
Northend
Ibstone
Cadmore End
Turville Cross
Turville Hth.
Skirmett
Fingest
Pishill
Stonor
Fawley
Hambleden
Bix
Middle Assendon
Lwr. Assendon
Aston
Remenham
Rotherfield Greys
Rotherfield Peppard
Cockpole Grn.
Harpsden
Lwr. Shiplake
Shiplake
Dunsden Grn.
Play Hatch
Sonning
Charvil

Bradenham
Naphill
West Wycombe
Beacon's Bottom
HIGH WYCOMBE
Lane End
Booker
Frieth
Mill End
Medmenham
Bisham
Hurley
Marlow
Lower Woodend
Lower Woodend
Rockwell End

Great Kingshill
Holmer Green
Hughenden Valley
Hazlemere
Downley
Wycombe Marsh
Penn
Handy Cross
Flackwell Heath
Little Marlow
Wooburn Green
Bourne End
Cookham
Cookham Dean
Cookham Rise
MAIDENHEAD
Warren Row
Knowl Hill
Littlewick Grn.
White Waltham
Woodlands Park
Bray
Burney

Amersham
Chenies
Little Chalfont
Chorleywood
Coleshill
RICKMANSWORTH
Penn St.
Winchmore Hill
Tylers Green
Chalfont St. Giles
Knotty Grn.
Seer Grn.
Chalfont Common
Horn Hill
Maple Cross
West Hyde
Chalfont St. Peter
Beaconsfield
Loudwater
Wooburn
Hedsor
Hedgerley
Gerrards Cross
Fulmer
Stoke Poges
Denham Green
Farnham Common
Taplow
Burnham
Farnham Royal
Britwell
Iver Heath
UXBRIDGE
Iver
SLOUGH
SLOUGH
Upton
Langley

Wargrave
Hare Hatch
Ruscombe
Twyford
St. Lawrence
Whistley Grn.
Woodley
Hurst
Thurlock Row
Paley St.
Holyport
WINDSOR & MAIDENHEAD
WINDSOR
Eton
Hawthorn Hill
Datchet
Horton
Old Windsor
Poyle
Wraysbury
Hythe End
EGHAM

hurst
READING
READING
Whitley
Burghfield
Threemile Cross
Grazeley
Wokefield Park
timer
Swallowfield
Beech Hill
Stratfield Saye
Earley
Winnersh
WOKINGHAM
Wokingham
Shinfield
Arborfield Cross
Barkham
Spencer's Wood
Arborfield Garrison
Farley Hill
Finchampstead
Riseley
Eversley
Binfield
Popeswood
Sindlesham
Arborfield
Bracknell
Easthampstead
Wick Hill
Crowthorne
Warfield
Newell Grn.
BRACKNELL
Bracknell
Winkfield
Winkfield Row
Ascot
Sunninghill
Bagshot
Cranbourne
Woodside
Sunningdale
Virginia Water
Lyne
Englefield Green
Thorpe
Addlestone
Windlesham
Burrowhill
Ottershaw
Chobham
Woodham

Stratfield Turgis
Turgis Grn.
Heckfield
Bramshill
Eversley Cross
Yateley
Sandhurst
Frogmore
Blackwater
Lightwater
Donkey Town
West End
Bisley
CAMBERLEY
Knaphill
Pyrford
WOKING

Hazeley
Hound Grn.
Hartfordbridge
Mattingley
Hartley Wespall
West Grn.
Rotherwick
Hook
Newnham
Basing
Hawley
Fox Lane
West Heath
Cove
FRIMLEY
Deepcut
Brookwood
Mytchett
Pirbright
Normandy
Mayford
Send
Worplesdon
Pitch Place
Stoughton
Burpham

Hartley Wintney
Phoenix Grn.
Winchfield
FLEET
Dogmersfield
North Warnborough
Crookham Village
Church Crookham
Odiham
ALDERSHOT
FARNBOROUGH
GUILDFORD
Wood Street
Ash

Up Nately
Greywell
edurwell
Tunworth
Upton Grey
Long Sutton
Weston Patrick
South Warnborough
Well
FARNHAM
Mill Lane
Crondall
Upper Hale
Runfold
Shackleford
Farncombe
Bramley

County boundary

District boundary

9

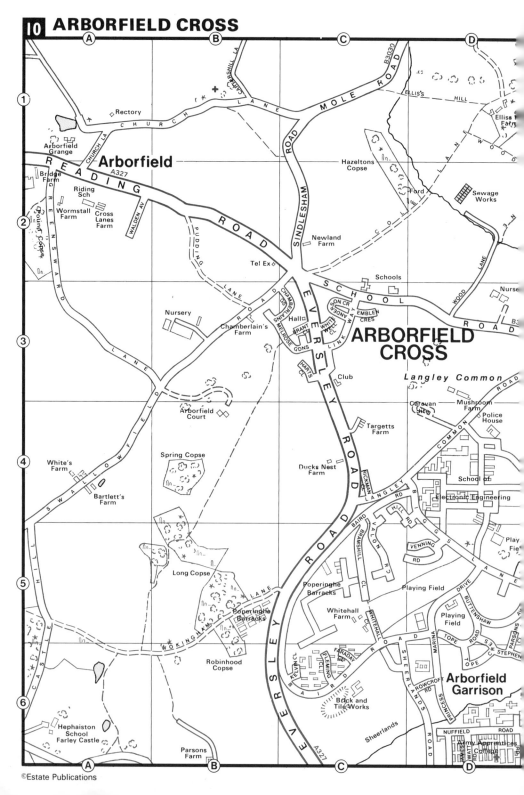

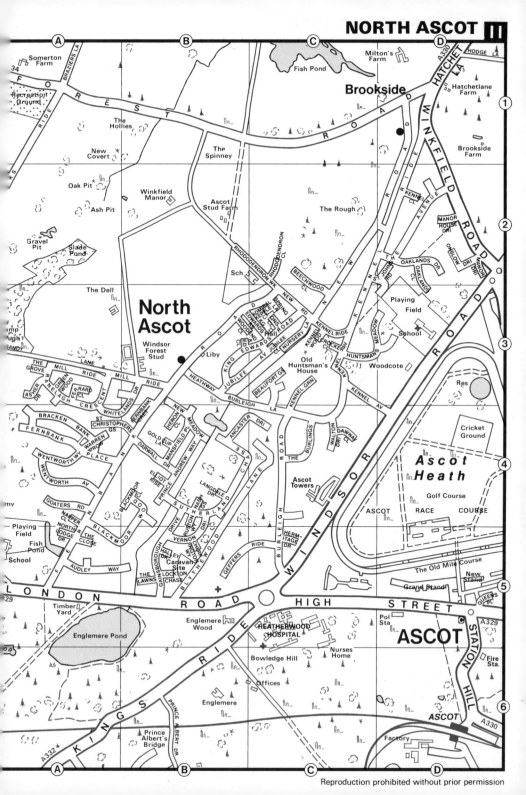

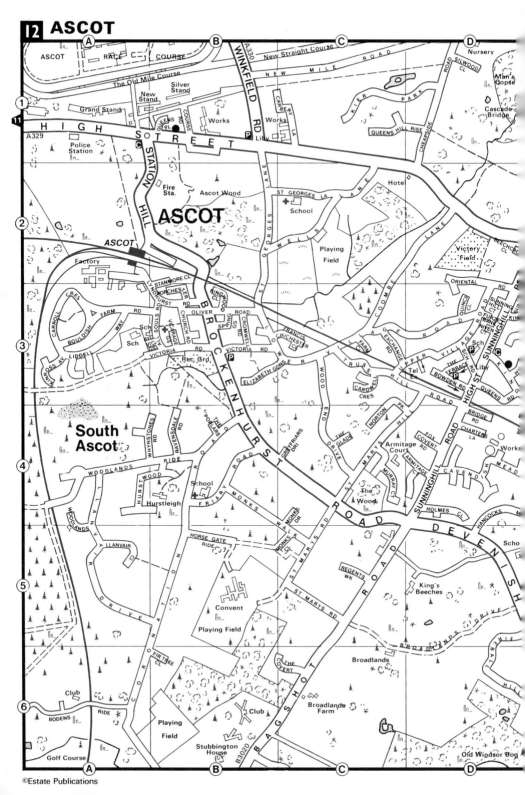

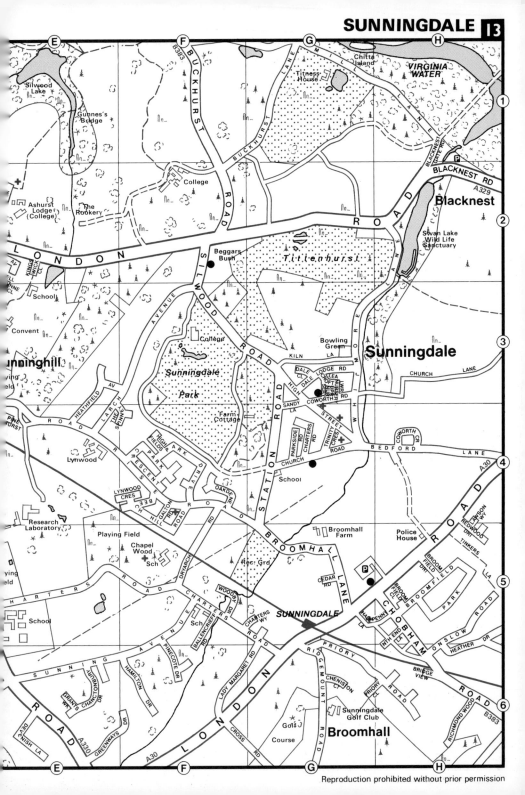

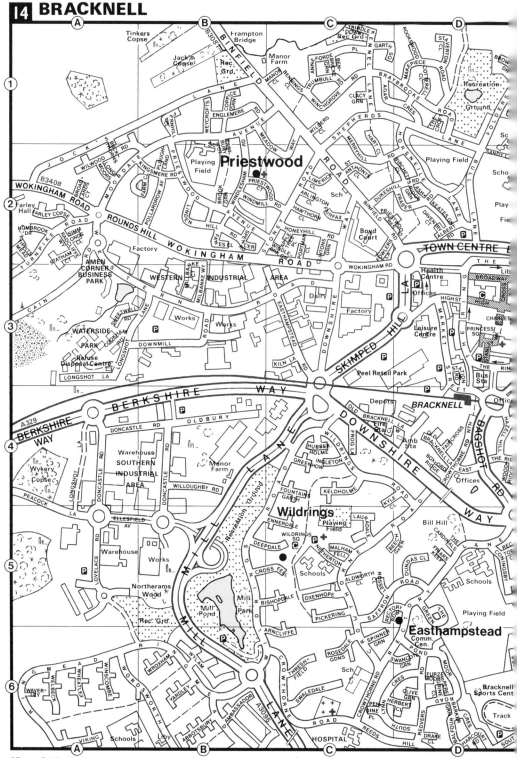

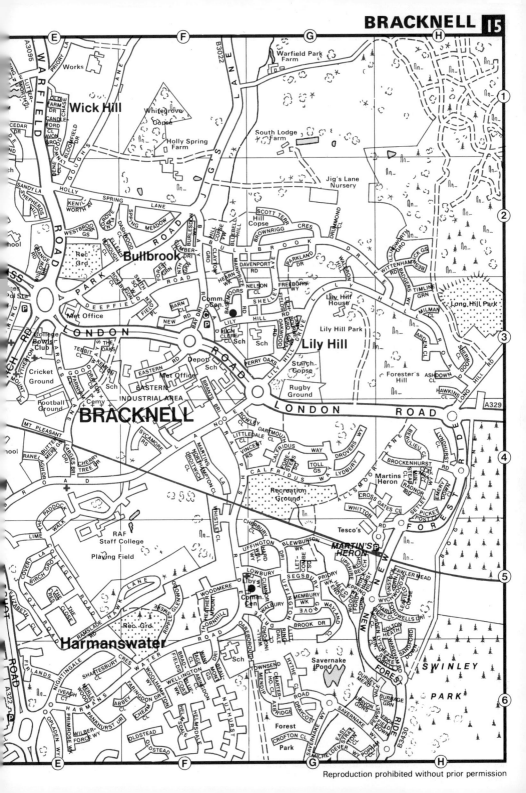

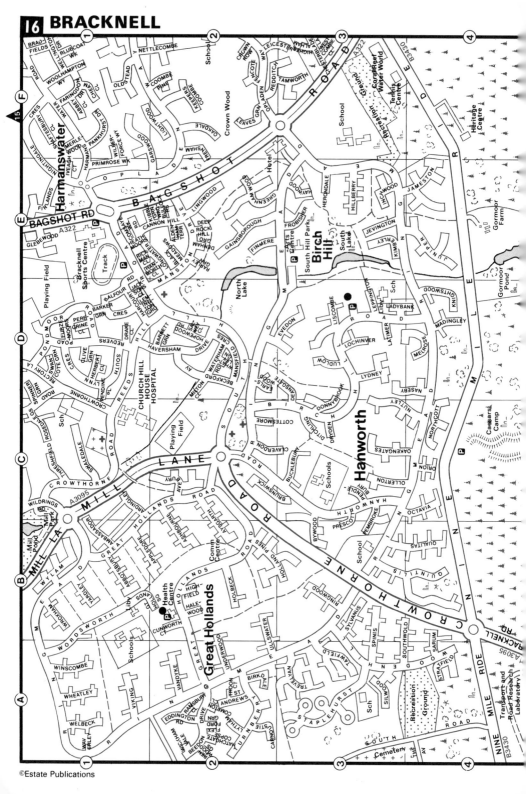

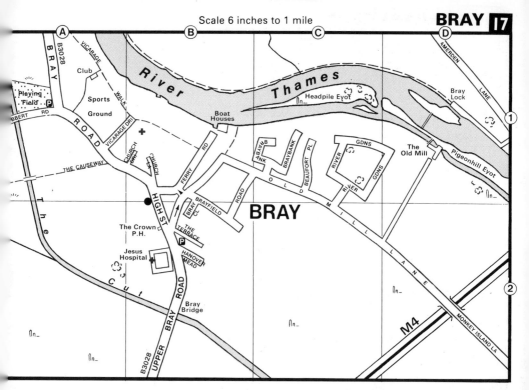

BRAY

BURGHFIELD COMMON

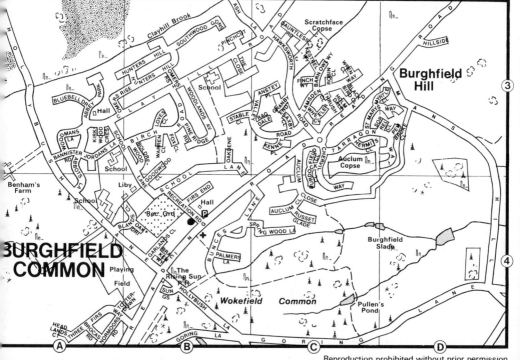

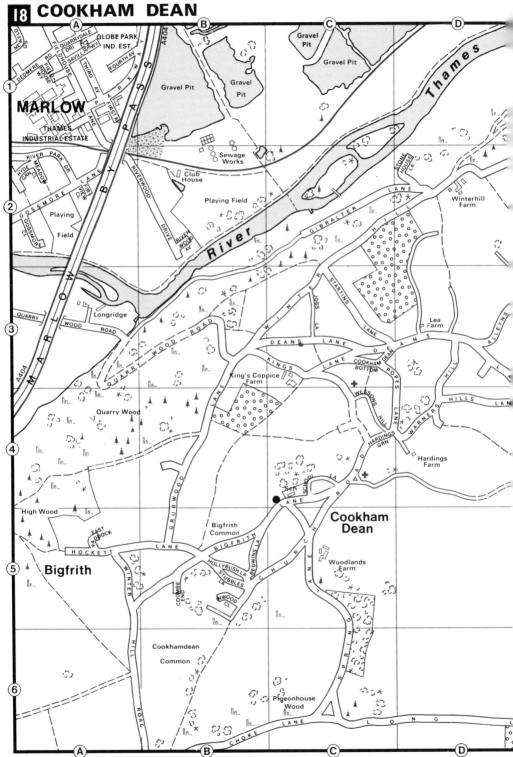

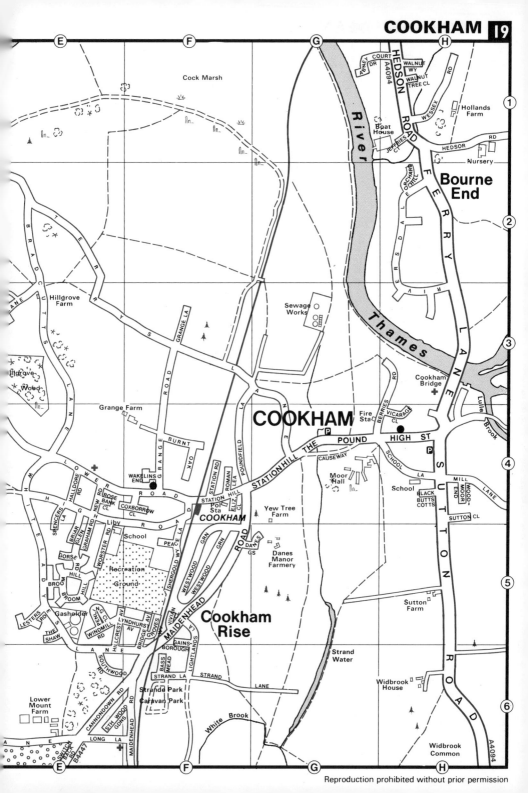

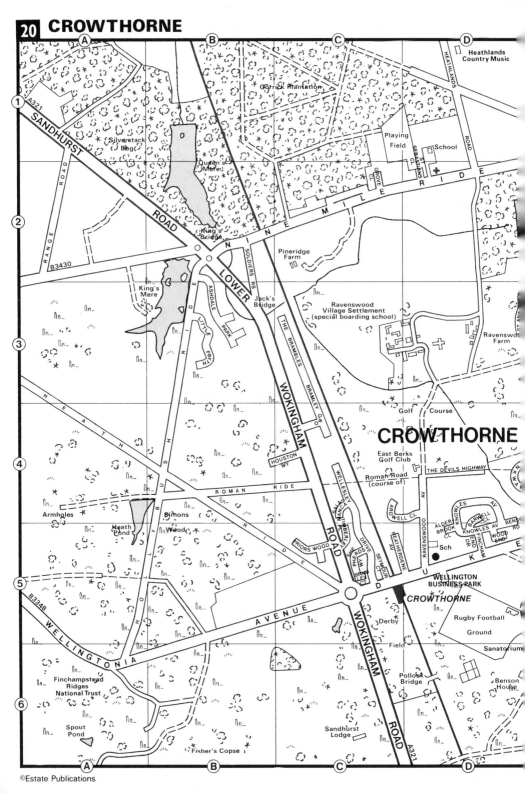

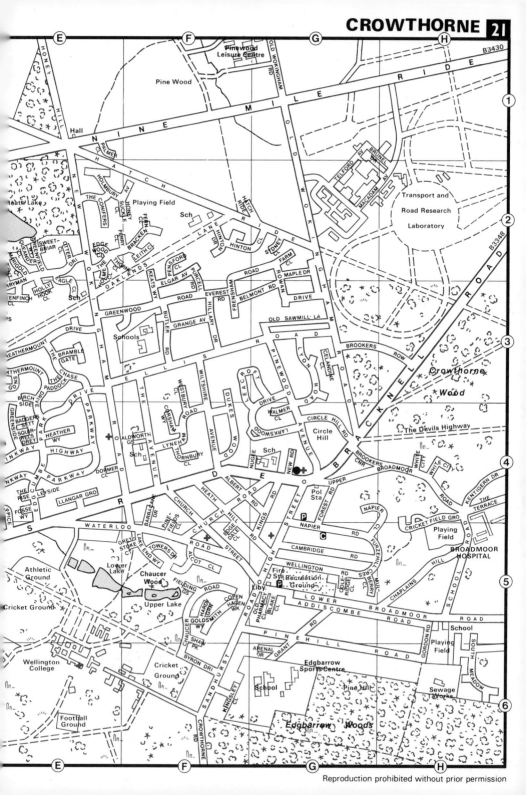

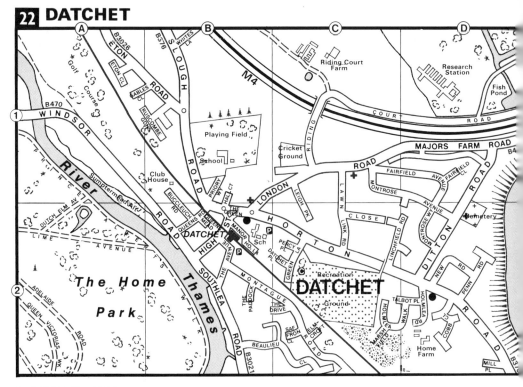

LAMBOURN

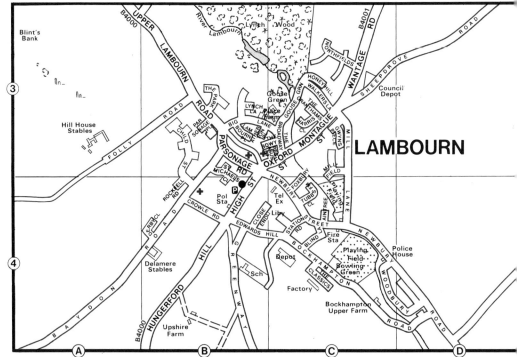

©Estate Publications

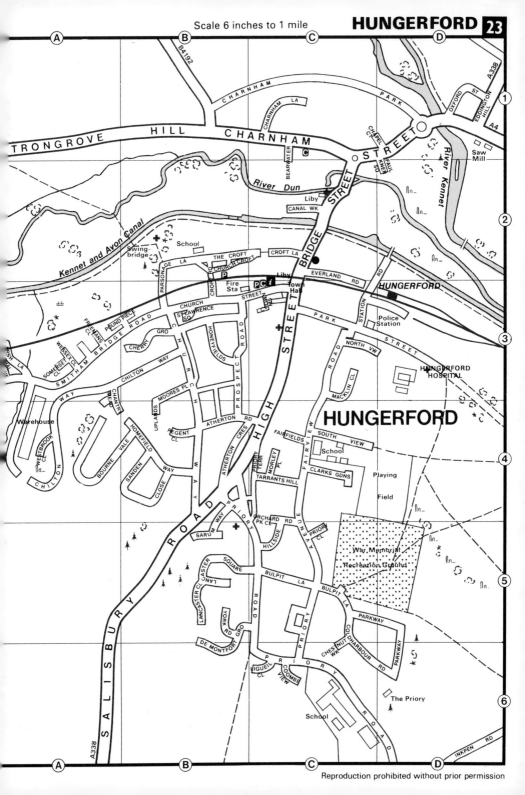

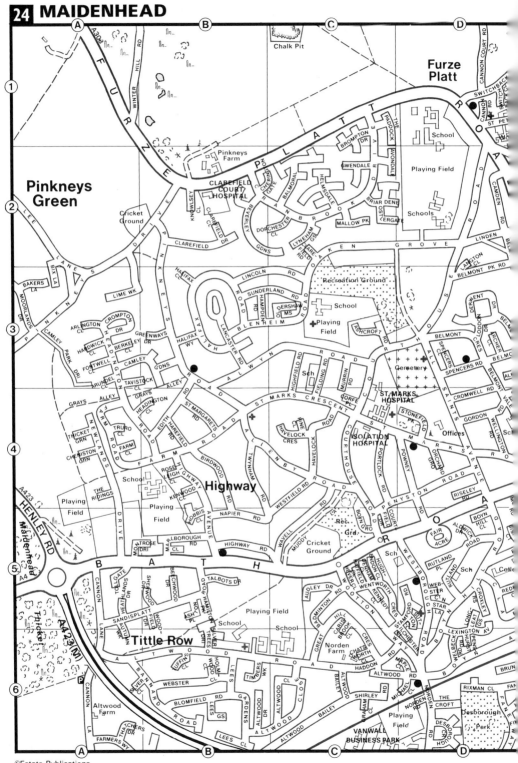

Furze Platt

Pinkneys Green

Highway

Tittle Row

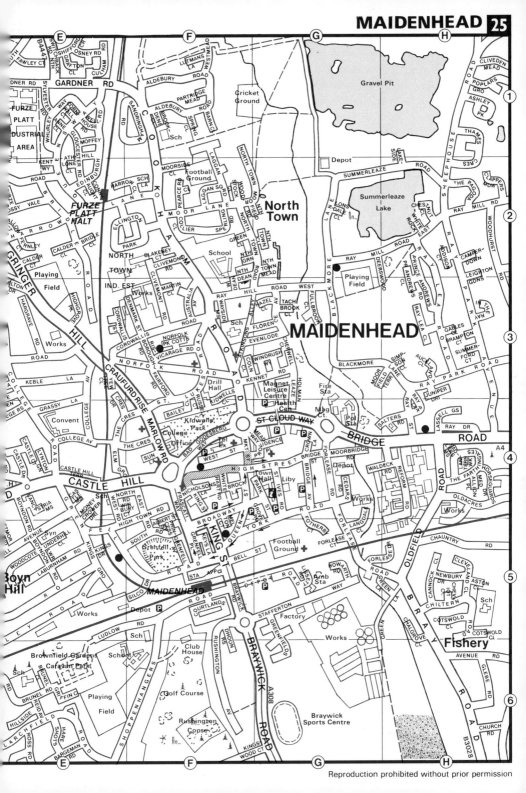

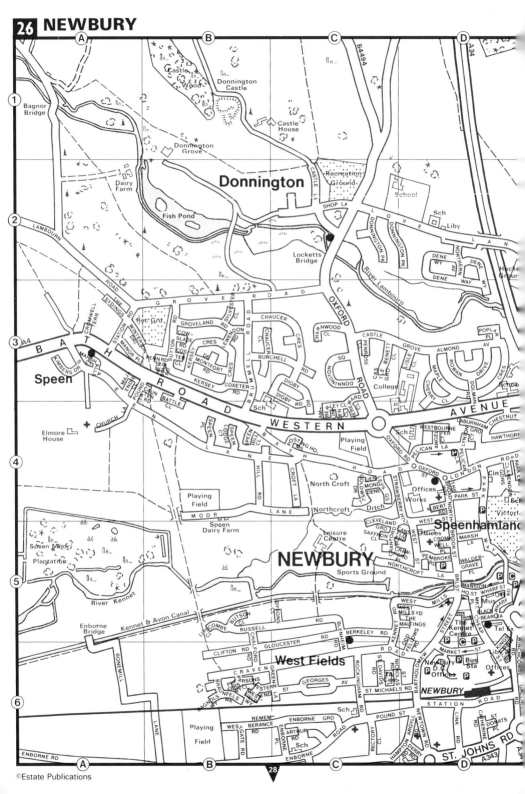

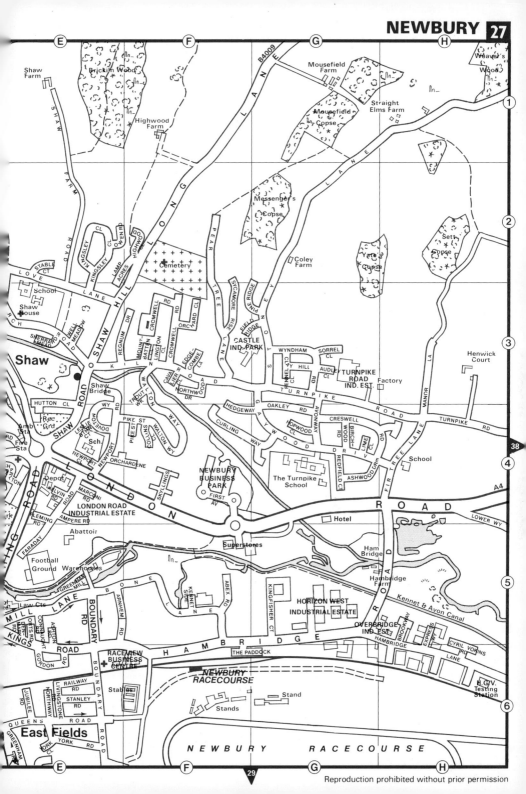

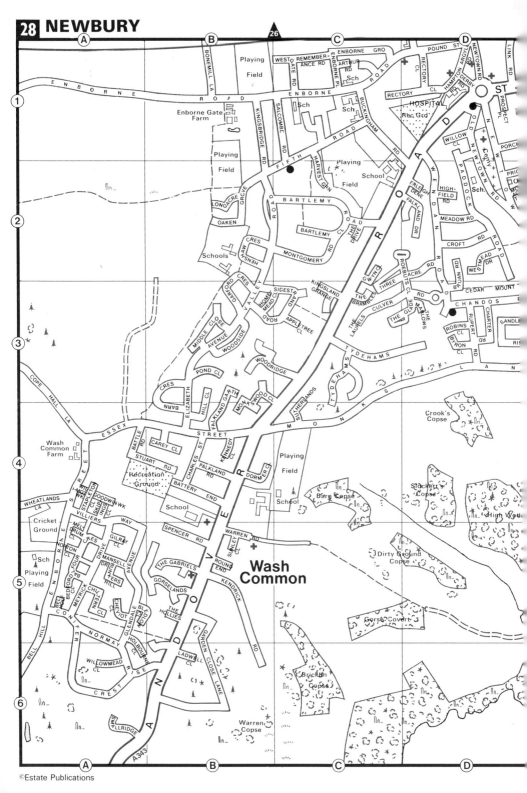

26

Wash Common

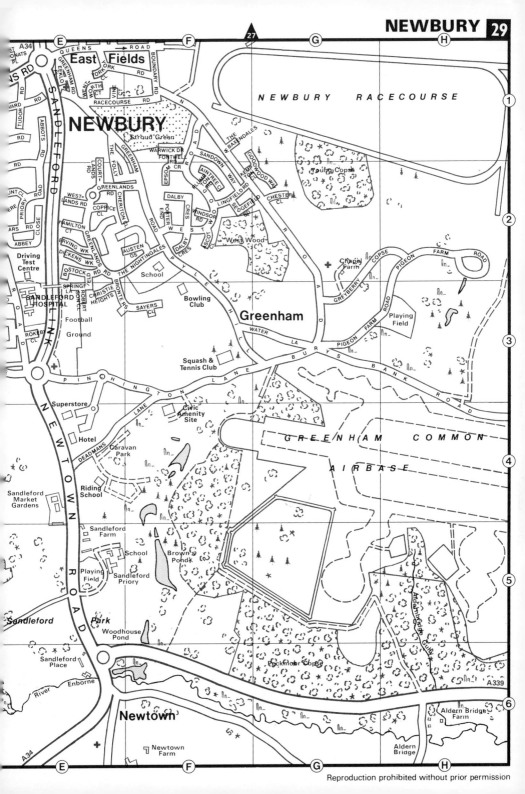

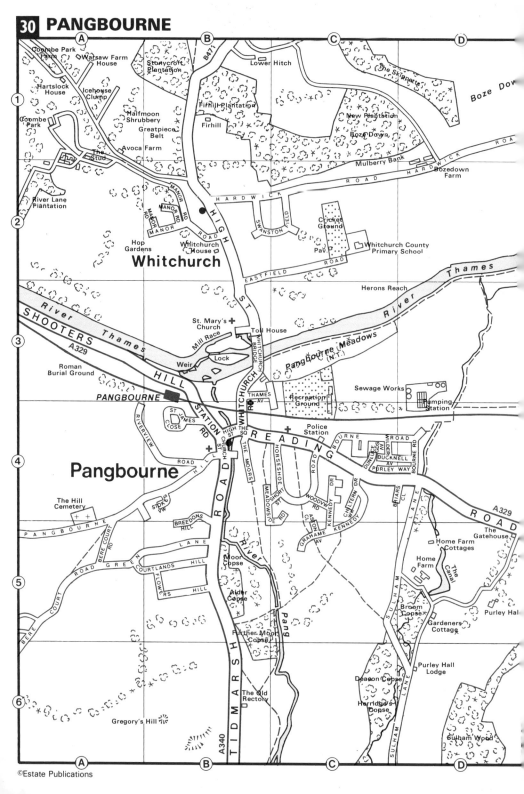

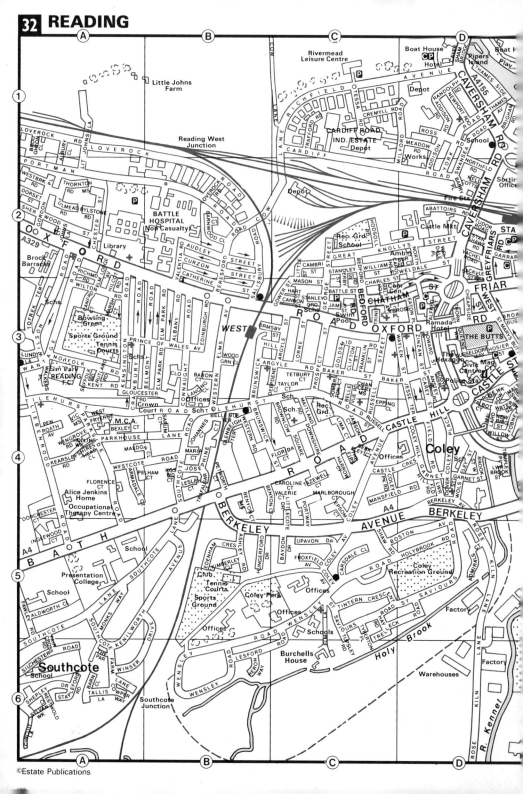

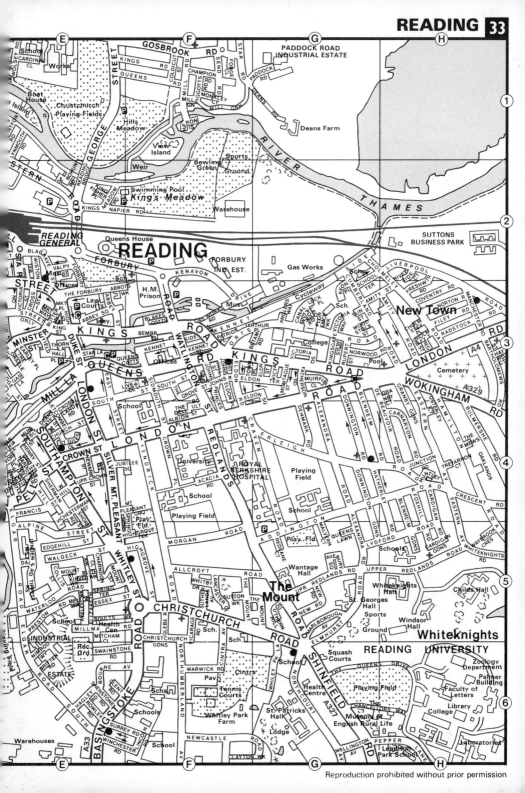

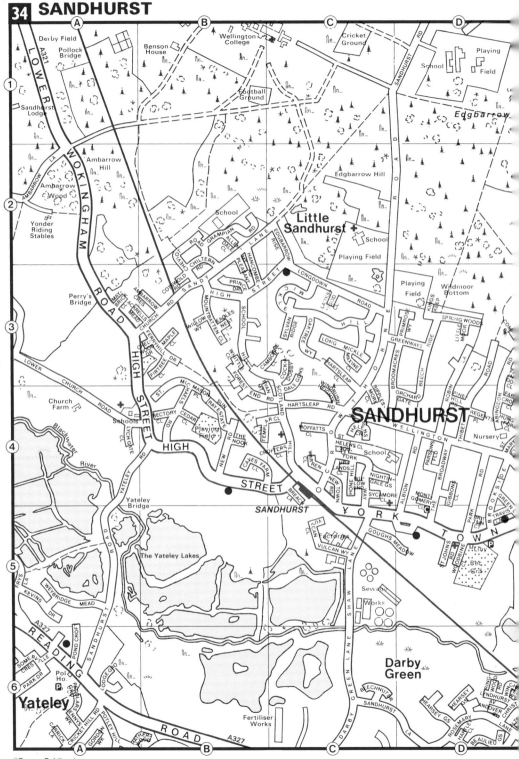

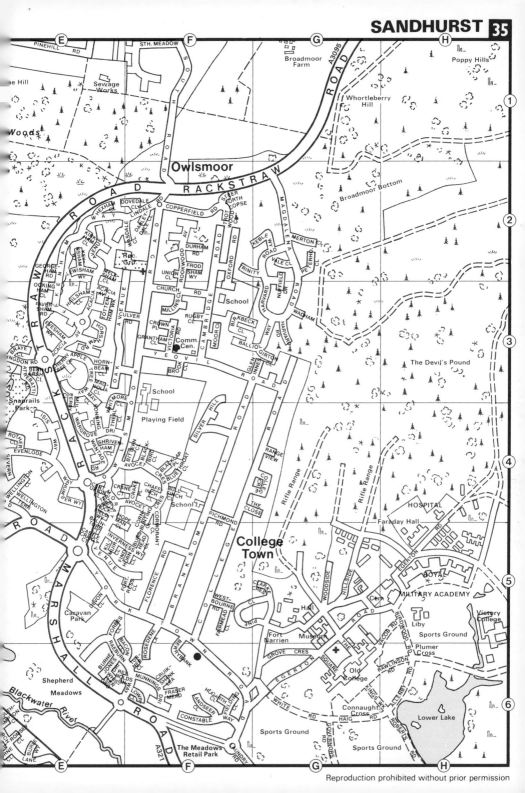

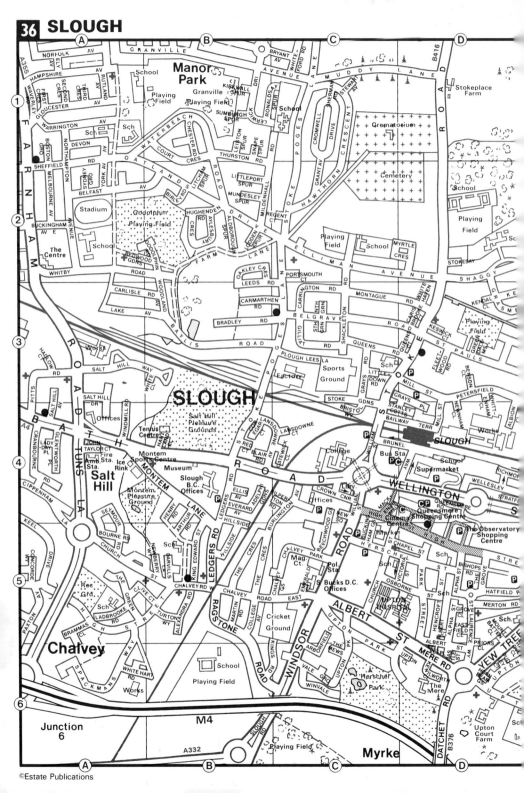

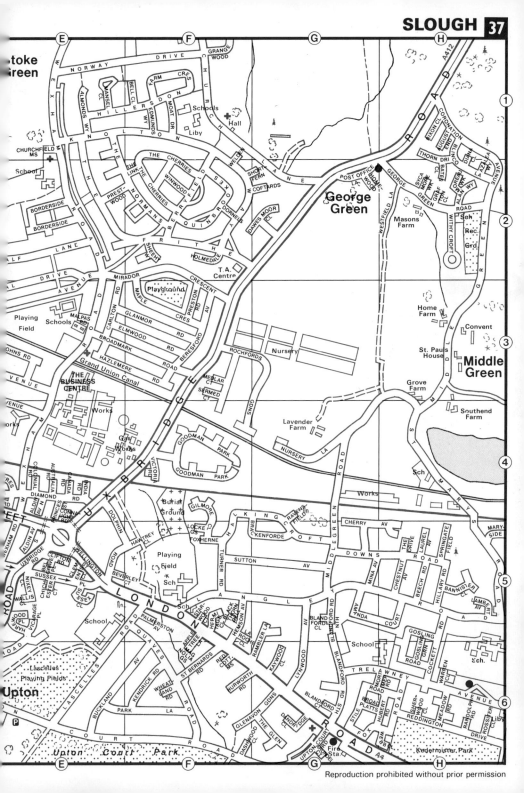

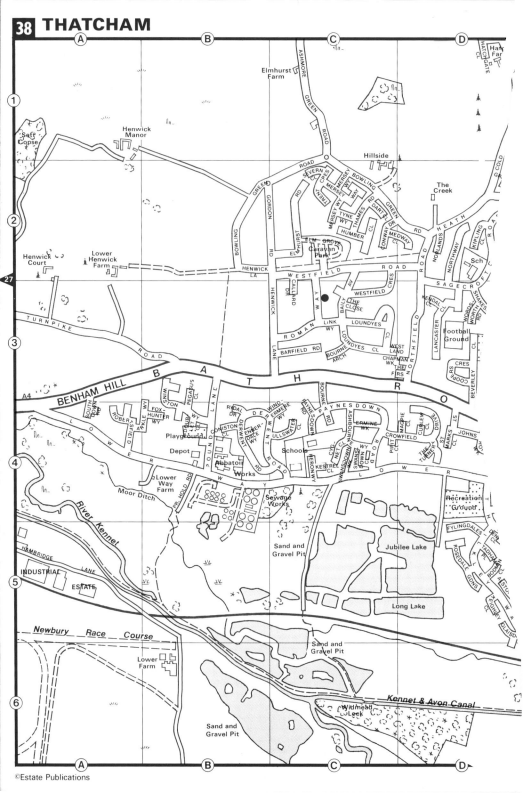

38 THATCHAM

©Estate Publications

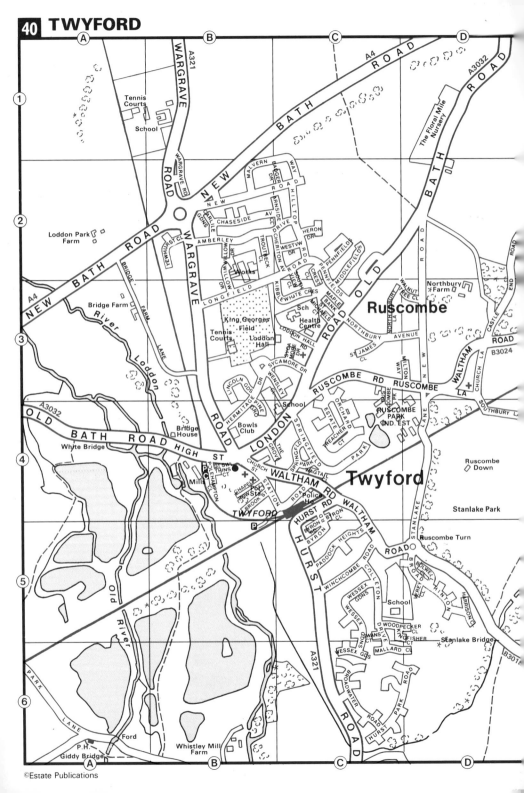

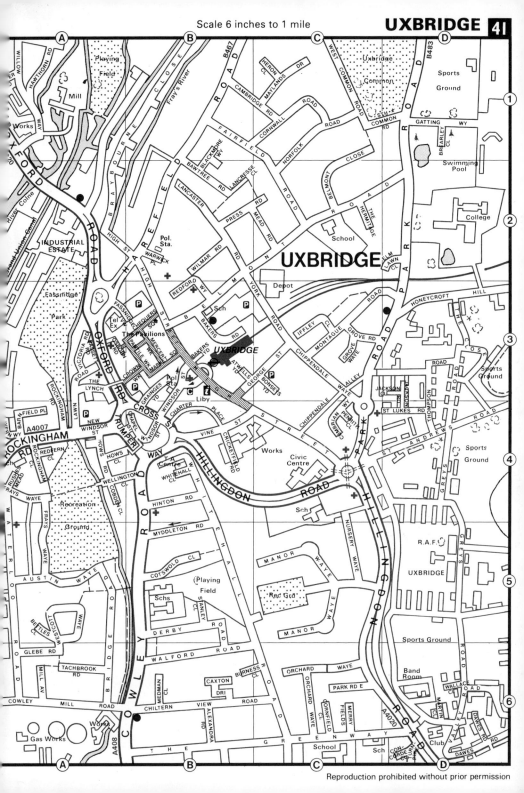

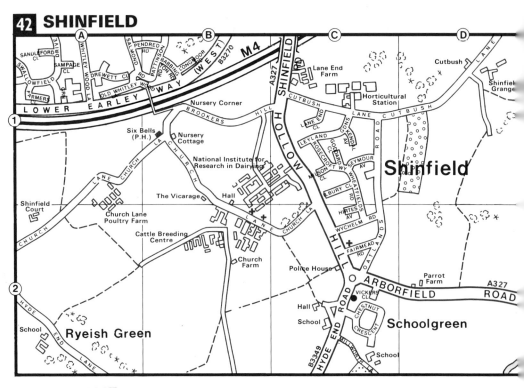

WARGRAVE

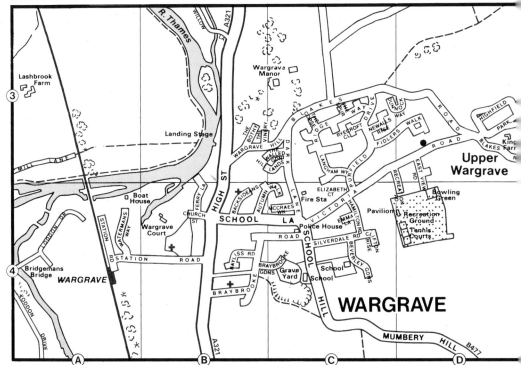

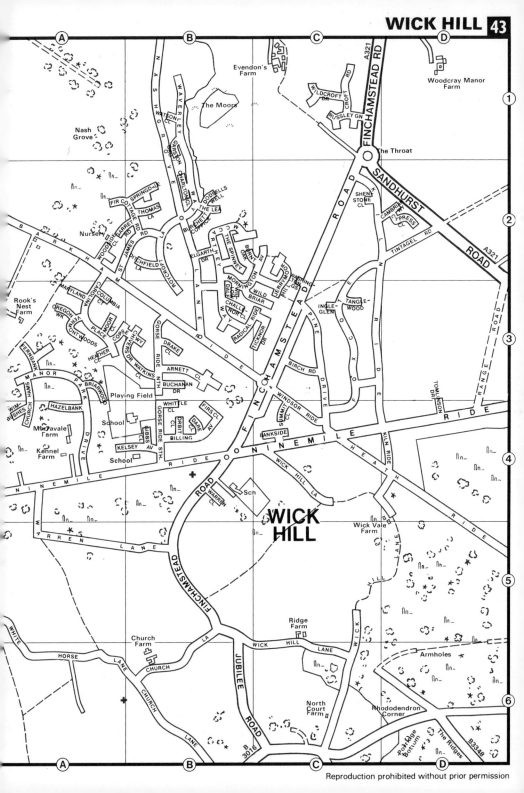

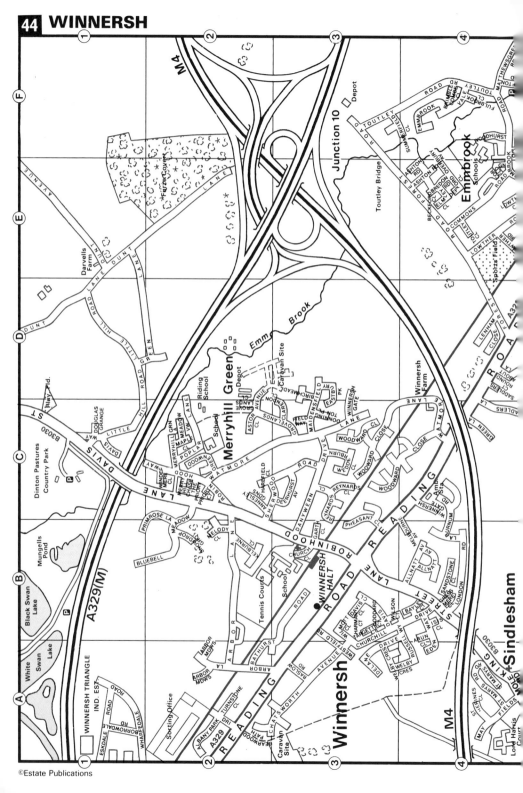

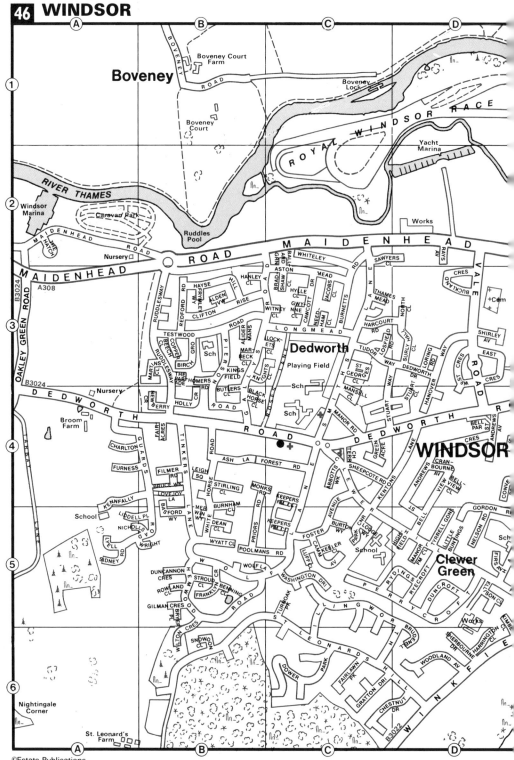

Boveney

Boveney Court Farm

Boveney Court

Boveney Lock

Yacht Marina

ROYAL WINDSOR RACE

RIVER THAMES

Windsor Marina

Caravan Park

Works

Ruddles Pool

MAIDENHEAD ROAD

LIME HATCH

Nursery

MAIDENHEAD ROAD

A308

B3024

OAKLEY GREEN ROAD

B3024

WHITELEY

SAWYERS CL

Dedworth

Playing Field

DEDWORTH ROAD

WINDSOR

Clewer Green

Nursery

Broom Farm

School

Kennfally

Nicholls

Liddell Pl

School

Nightingale Corner

St. Leonard's Farm

B3022

©Estate Publications

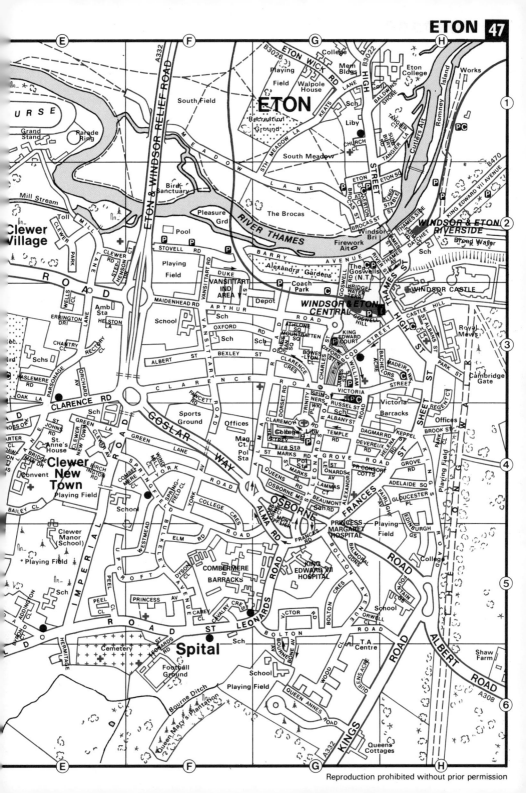

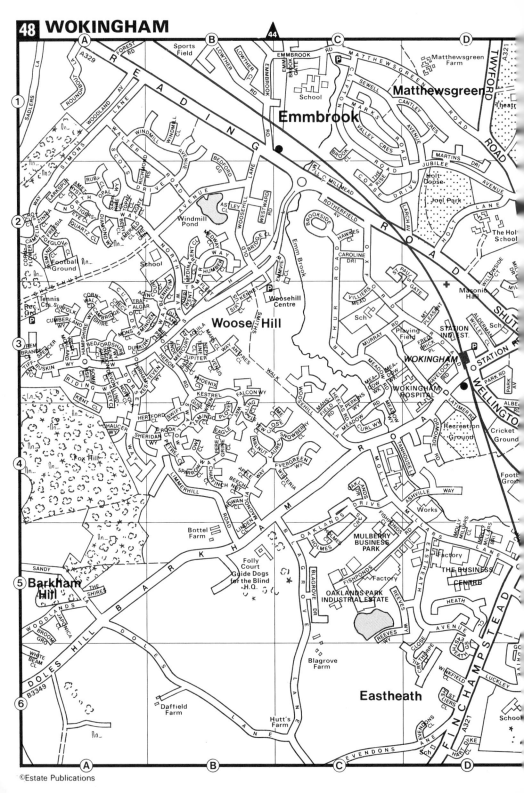

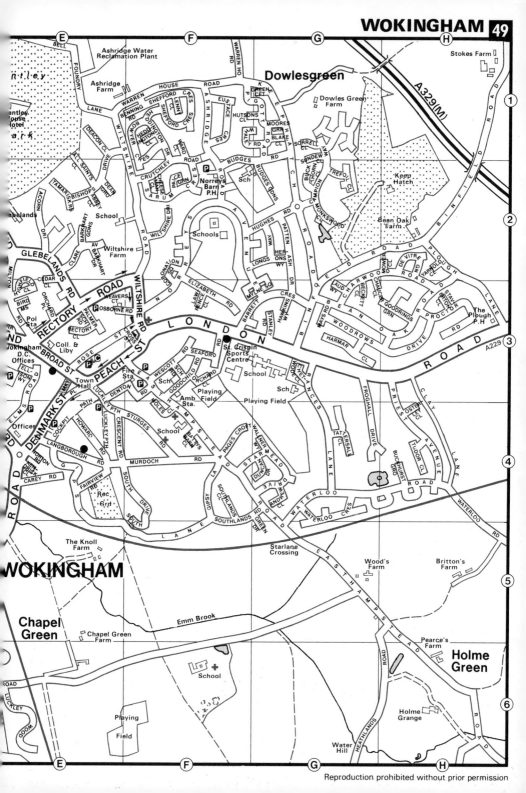

INDEX TO STREETS

ARBORFIELD CROSS

Anderson Cres	10 C3
Baird Rd	10 C5
Baird Rd	10 C6
Biggs La	10 D4
Bramshill Clo	10 C5
Brant Clo	10 C3
Buttenshaw Av	10 D5
Cartershill La	10 B1
Castle Hill	10 A6
Chamberlains Gdns	10 C3
Church La	10 A2
Cole La	10 C2
Ellis's Hill	10 C1
Emblen Cres	10 C3
Eversley Rd	10 C3
Faraday Clo	10 C6
Fleming Clo	10 C6
Gravelpithill La	10 D1
Greensward La	10 A2
Harts Clo	10 C3
Hilbron Rd	10 D6
Hill Rd	10 C4
James Watt Rd	10 D6
Kelvin Clo	10 C6
Langley Common Rd	10 C4
Link Way	10 C3
Melrose Gdns	10 B3
Mole Rd	10 C1
Nuffield Rd	10 D6
Parsons Clo	10 D6
Princess Marina Dri	10 D6
Pudding La	10 B2
Reading Rd	10 A2
Rickman Clo	10 C4
Rowcroft Rd	10 D6
School Rd	10 C3
Sheerlands Rd	10 C6
Sindlesham Rd	10 C2
Stephenson Rd	10 D6
Swallowfield Rd	10 A4
Tope Cres	10 D6
Tope Rd	10 D5
Valon Rd	10 C5
Venning Rd	10 D5
Walden Av	10 A2
Whitehall Clo	10 C5
Whitewell Clo	10 C3
Wokingham La	10 B6
Wood La	10 D3

ASCOT

Admiral Kepple Ct	11 C3
All Souls Rd	12 B3
Ancaster Dri	11 B4
Armitage Ct	12 D4
Asher Dri	11 A3
Audley Way	11 A5
Bagshot Rd	12 B6
Ballencriefg Rd	13 F5
Beaufort Gdns	11 C3
Bedford La	13 G4
Beech Hill Rd	13 F4
Beechcroft Clo	12 D2
Beechwood Clo	11 C3
Blackmoor Clo	11 B4
Blackmoor Wood	11 A5
Blacknest Gate Rd	13 H1
Blacknest Rd	13 H2
Blythewood La	11 B5
Bodens Ride	12 A6
Bouldish Farm Rd	12 A3

Bowden Rd	12 D3
Bracken Bank	11 A4
Braziers La	11 A1
Bridge Rd	12 D4
Bridge Vrw	13 H6
Broadlands Dri	12 C5
Brockenhurst Rd	12 B3
Broomfield Clo	13 H5
Broomfield Dri	13 H5
Broomfield Pk	13 H5
Broomhall La	13 G5
Buckhurst La	13 F1
Buckhurst Rd	13 F1
Burleigh La	11 B3
Burleigh Rd	11 C5
Carbery La	12 C1
Cardwell Cres	12 C3
Carroll Cres	12 A3
Cavendish Meads	12 D4
Cedar Rd	13 G5
Chanctonbury Dri	13 E6
Charters La	12 D4
Charters Rd, Sunningdale	13 G4
Charters Rd, Sunninghill	13 E5
Charters Way	13 G3
Cheapside Rd	12 D1
Cheniston Ct	13 G6
Cherrington Way	11 B5
Chobham Rd	13 H5
Church La, Sunningdale	13 H3
Church La, Sunninghill	13 E2
Church Rd, Ascot	12 B3
Church Rd, Sunningdale	13 G4
Coombe La	12 C3
Coronation Rd	12 A6
Course Rd	12 B1
Coworth Cres	13 H4
Coworth Rd	13 G3
Crocker Clo	11 C3
Cromwell Rd	12 B3
Cross Rd	13 F6
Dale Clo	13 G3
Dale Lodge Rd	13 G3
Darwall Dri	11 B4
Dawnay Clo	11 C4
Devenish La	13 E6
Devenish Rd	12 D5
Dryarch Rd	13 F5
Elizabeth Gdns	12 B3
Elliot Rise	11 B4
Exchange Rd	12 C3
Farm Clo	12 C3
Fernbank Cres	11 B4
Fernbank Pl	11 A4
Fernbank Rd	11 A5
Ferrard Clo	11 A3
Fir Tree Clo	12 B6
Fireball Hill	12 D5
Forest Rd	11 A1
Fox Covert Rd	12 D4
Francis Chichester Clo	12 C3
Friary Rd	12 B4
Furzebank	12 D3
Gainsborough Dri	11 B5
Galton Rd	13 F4
Geffers Ride	11 B5
Goaters Rd	11 A4
Gold Cup La	11 B4
Grants Walk	13 E6
Green Wood	11 A3
Greenways Dri	13 E6
Greyfriars Dri	12 C4
Halfpenny La	13 G5
Halley Dri	11 B5
Hamilton Dri	13 F6
Hancocks Mt	12 D5
Hatchet La	11 D1
Heather Dri	13 H6
Heathfield Av	13 E4
Heathway	11 B3
Hermitage Dri	11 C5
Heron Clo	11 B4
High Fields	13 F4
High St, Ascot	11 C5

High St, Sunningdale	13 G3
High St, Sunninghill	12 D3
Highclere	12 D3
Hodge La	11 D1
Holmes Clo	12 D4
Horse Gate Ride	12 B5
Huntsmans Meadow	11 C3
Hurstwood	12 A4
Jubilee Av	11 B3
Jubilee Clo	11 C3
Kennel Av	11 C3
Kennel Clo	11 D2
Kennel Grn	11 C3
Kennel Ride	11 C3
Kennel Wood	11 C3
Kier Pk	12 C1
Kiln La	13 G3
King Edwards Clo	11 B3
King Edwards Rise	11 C3
King Edwards Rd	11 B3
Kings Ride	11 A6
Kings Rd	12 D3
Kingswick Clo	13 E2
Kingswick Dri	12 D3
Kinross Av	12 A3
Lady Margaret Rd	13 F6
Langdale Dri	11 B4
Larch Av	13 E4
Lawson Way	13 H4
Lea Croft	13 G3
Liddell Way	12 A3
Llanvair Clo	12 A5
Llanvair Dri	12 A4
Locks Ride	11 A2
Lockton Chase	11 B5
London Rd, Burleigh	11 A5
London Rd, Sunningdale	13 F6
London Rd, Sunninghill	13 E2
Lower Nursery	13 G2
Lower Village Rd	12 B3
Lyndhurst Rd	12 A3
Lynwood Cres	13 F4
Manor Dri	11 D2
Manor House Dri	11 D2
Mansfield Pl	11 B4
Mill La	13 G1
Mill Ride	11 A3
Monks Clo	12 C5
Monks Dri	12 C4
Monks Walk	12 B4
Murray Ct	12 C4
Napper Clo	11 A4
Nash Gdns	11 B4
Nell Gwynne Av	13 E2
Nell Gwynne Clo	13 E2
New Meadow	11 B4
New Mile Rd	12 B1
New Rd	11 C3
Niven Ct	12 D3
North End La	13 H5
North Lodge Dri	11 A5
Norton Pk	12 C4
Nursery La	11 C3
Oakdene	13 F4
Oaklands Clo	11 D2
Oaklands Dri	11 D2
Oliver Rd	12 B3
Onslow Rd	13 H6
Onslow Dri	11 D2
Oriental Rd	12 D2
Park Cres	13 F4
Park Dri	13 F4
Parkside Rd	13 G4
Pembroke Clo	12 D4
Pinecote Dri	13 F6
Pinehurst	13 E4
Porchester	12 B3
Prince Albert Dri	11 B6
Prince Andrew Way	11 B4
Priory Clo	13 G6
Priory Rd	13 G6
Queens Clo	11 C3
Queens Hill Rise	12 D1
Queens Pl	11 D5

Queens Rd	12 D3
Quince Clo	12 D3
Ranelagh Cres	11 A3
Ravensdale Rd	12 B4
Redwood Dri	13 H5
Regents Walk	12 C5
Rhododendron Clo	11 C2
Rhododendron Walk	11 B2
Richmond Wood	13 H6
Ridgemount Rd	13 G6
Ringwood Clo	12 B3
Rise Rd	13 E4
Royal Victoria Gdns	12 A3
Ruston Way	11 B4
St Christophers Gdns	11 A4
St Georges La	12 B3
St Johns Rd	11 D2
St Marys Hill	12 C4
St Marys Rd	12 C5
Sandy La, North Ascot	11 A3
Sandy La, Sunningdale	13 G4
School Rd	12 D3
Silwood Clo	12 D1
Silwood Rd	13 F2
Spring Gdns	12 B3
Spring La	11 C3
Stanmore Clo	12 B2
Station Hill	11 D5
Station Rd	13 G4
Sunning Av	13 E6
Sunninghill Clo	12 D3
Sunninghill Ct	12 D3
Sunninghill Rd	12 D3
Sutherland Chase	11 B4
Tenby Dri	13 E4
The Avenue	11 D2
The Burlings	11 C4
The Close	11 A5
The Covert	12 C6
The Glade	12 C4
The Grove	11 A3
The Lawns	11 B5
The Poplars	12 B4
The Spinney	13 E4
The Terrace	12 D3
Tinkers La	13 H5
Trinity Cres	13 G4
Truss Hill Rd	12 C3
Upper Nursery	13 G3
Upper Village Rd	12 C3
Vernon Dri	11 B5
Vicarage Gdns	12 B3
Victoria Rd	12 B3
Walton Dri	11 C4
Warren Row	11 A4
Wells La	12 C2
Wentworth Av	11 A4
Wentworth Way	11 A4
Whitelands Dri	11 A4
Whitmore La	13 G4
Whynstones Rd	12 A4
Windsor Rd	11 C5
Winkfield Rd	11 D1
Wood End Rd	12 C3
Woodby Dri	13 F5
Woodend Clo	11 B3
Woodlands Clo	12 A5
Woodlands Ride	12 A4

BRACKNELL

Abbey Clo	15 F6
Abbotsbury	16 B1
Abingdon Clo	15 F6
Agar Cres	14 C1
Albert Rd	14 D2
Aldenham Ter	16 E2
Aldworth Clo	14 C5
Allsmoor La	15 G4
Ambassador	16 B1
Anneforde Pl	14 C1
Appledore	16 B1
Appletree Pl	14 C2
Arden Clo	15 H3
Ardingley	16 C1
Arlington Clo	14 C2
Arncliffe	14 B6

Ashbourne	16 B2	
Ashdown Clo	15 H3	
Ashridge Grn	14 D2	
Avebury	16 C2	
Avon Gro	15 E1	
Axbridge	15 G6	
Aysgarth	16 B2	
Badgers Way	15 H3	
Bagshot Rd	14 D4	
Balfour Cres	16 D1	
Barker Grn	16 D1	
Barn Clo	15 F3	
Barnett Grn	16 D2	
Basemoors	15 F3	
Bay Dri	15 F3	
Bay Rd	15 F2	
Beaulieu Clo	15 H4	
Beaumont Gdns	15 F6	
Beckford Av	16 D2	
Beech Glen	14 D5	
Beedon Dri	16 A2	
Benbricke Grn	14 C1	
Bennings Clo	14 C1	
Berkshire Way	14 A4	
Beswick Gdns	15 H2	
Binfield Rd	14 B1	
Birch Gro	15 E5	
Birch Hill Rd	16 C2	
Birchetts Clo	14 D2	
Birkdale	16 A2	
Bishopdale	14 B5	
Black Meadows	16 E2	
Blewburton Walk	15 G5	
Bloomfield Dri	15 E1	
Blue Coat Walk	15 F6	
Bluebell Hill	15 F2	
Boxford Ridge	14 D4	
Bracknell Rd	16 A4	
Bradfields	15 F6	
Brants Bri	15 F4	
Braybrooke Rd	14 C1	
Broad La	15 E4	
Broadlands Ct	14 A2	
Broadway	14 D3	
Brockenhurst Rd	15 H4	
Brook Dri	15 G5	
Brook Grn	14 B2	
Brownlow Dri	14 D1	
Brownrigg Cres	15 G2	
Brunswick	16 C2	
Buckhurst Hill	15 G5	
Bucklebury	16 C3	
Buckthorns	14 A2	
Bull La	14 D2	
Bullbrook Dri	15 F2	
Burbage Grn	15 H6	
Burlesdon Way	15 F2	
Burnham Gro	15 E1	
Buttermere Gdns	15 E4	
Bywood	16 B3	
Cain Rd	14 A3	
Calfridus Way	15 G4	
Candleford Clo	15 E1	
Cannon Hill	16 E1	
Carnoustie	16 A2	
Cedar Dri	15 E1	
Charles Sq	14 D3	
Charterhouse Clo	15 G6	
Cheam Clo	15 F6	
Cherbury Clo	15 G5	
Cherry Tree Dri	15 E4	
Chesterblade La	15 F3	
Church Rd	15 E3	
Clacy Grn	14 C1	
Claverdon	16 C2	
Clayhill Clo	15 H4	
Clayton Gro	15 F2	
Clintons Grn	14 C2	
Clive Grn	14 D6	
College Rd	15 E5	
Compton Clo	16 A2	
Comsaye Walk	16 E1	
Coningsby	14 D5	
Cookham Rd	14 A3	
Coombe Pine	16 F2	
Coppice Grn	14 B1	
Cotterell Clo	14 D1	
Cottesmore	16 C2	
Covert La	15 E5	
Crofton Clo	15 G6	
Cross Fell	14 B5	
Cross Gates Clo	15 H4	
Crossway	14 D3	
Crown Row	16 F2	
Crowthorne Rd, Easthampstead	14 C6	
Crowthorne Rd, Hanworth	16 B4	
Crowthorne Rd Nth	14 D4	
Cumberland Dri	15 F2	
Cumnor Way	15 G5	
Cunworth Ct	16 B2	
Dartmouth Clo	15 G4	
Dashwood Clo	15 F2	
Davenport Rd	15 F2	
Daventry Ct	14 D2	
Deansgate	16 D3	
Deep Rock Hill	16 E2	
Deepdale	14 B5	
Deepfield Rd	15 E3	
Dene Clo	15 E2	
Denham Gro	16 E2	
Ditchling	16 C3	
Doncastle Rd	14 A4	
Donnybrook	16 C3	
Downmill Rd	14 B3	
Downshire Way	14 C4	
Drake Clo	16 D1	
Draycott	15 G6	
Drayton Clo	15 E3	
Droitwich Clo	15 F4	
Drovers Way	15 G4	
Drummond Clo	15 G2	
Dryden	16 C3	
Dukeshill Rd	14 C2	
Dundas Clo	14 D5	
Durley Mead	15 G6	
Earlswood	16 D2	
East Stratton Clo	15 G6	
Eastbury Ct	14 B2	
Eastern Ind Area	15 F3	
Eastern Rd	15 F3	
Easthampstead Rd	14 C3	
Eddington Rd	16 A2	
Elizabeth Clo	15 E5	
Ellenborough Clo	15 E2	
Ellesfield Av	14 A5	
Emerydown Clo	15 H4	
Englemere Rd	14 B1	
Ennerdale	14 C5	
Epping Way	15 G5	
Evedon	16 D3	
Faircross	14 D4	
Fairfax	14 C2	
Fanes Clo	14 B2	
Faringdon Dri	15 F6	
Farley Copse	14 A2	
Farm Clo	14 B2	
Farnham Clo	15 E3	
Fawler Mead	15 H5	
Fencote	16 F2	
Fernhill Clo	14 B1	
Fielden Pl	15 F3	
Finmere	16 E2	
Finstock Grn	15 H5	
Firlands	15 E6	
Flexford Grn	16 A2	
Flintgrove	15 E2	
Folders La	14 D1	
Fordwells Dri	15 H5	
Forest Grn	15 F2	
Foresters Sq	15 G4	
Fountains Garth	14 C4	
Fowlers La	14 C2	
Fox Rd	15 E5	
Fraser Rd	14 D2	
Freeborn Way	15 G3	
Frensham	16 F2	
Friars Keep	14 D5	
Frobisher	15 E2	
Frog La	14 C4	
Froxfield Down	15 H6	
Furze Moors	14 D6	
Gainsborough	16 E2	
Garswood	16 E2	
Garth Sq	14 C1	
Glebewood	14 D6	
Glenwood	15 F5	
Goodways	15 E3	
Gough's La	15 E2	
Grange Rd	15 E2	
Great Hollands Rd	16 A2	
Great Hollands Sq	16 B1	
Greenham Wood	16 E2	
Greenhow	14 C4	
Hale End	15 G5	
Halewood	16 B2	
Hanworth Clo	16 E1	
Hanworth Rd	16 C3	
Harcourt Rd	16 D1	
Hardwell Way	15 G5	
Harlbury Clo	15 H5	
Harmans Water Rd	15 E6	
Hart Clo	14 C2	
Hatchgate Copse	16 A2	
Haversham Dri	16 D1	
Hawk La	15 E5	
Hawkins Clo	15 H3	
Hawthorn Clo	14 C2	
Haywood	16 E3	
Hazell Hill	14 D4	
Hearn Walk	15 F2	
Heath Moors	15 E1	
Heathermount	15 F5	
Helmsdale	15 F6	
Herbert Clo	14 C6	
Herondale	16 E3	
High Beech	15 G5	
High St	14 D3	
Highclere Clo	15 F3	
Higher Alham	15 F3	
Highfield	16 B2	
Hill Copse Vw	15 F2	
Hillberry	16 E3	
Holbeck	16 B2	
Holland Pines	16 B2	
Holly Spring La	15 E2	
Holton Heath	15 H5	
Hombrook Dri	14 A2	
Honeyhill Rd	14 C2	
Hopewood Rd	16 D2	
Horsneile La	14 D1	
Hubberholme	14 C4	
Hythe Clo	15 G6	
Illingworth Gro	15 H2	
Inchwood	16 E3	
Ingleton	14 C4	
Iveagh Ct	15 E6	
Jacksons Clo	16 D1	
Jameston	16 E4	
Jevington	16 E3	
Jig's La	15 F2	
Jocks La	14 A2	
Juniper	16 E4	
Keates Grn	14 D2	
Keepers Coombe	16 F2	
Keldholme	14 C4	
Kenilworth Av	15 E2	
Kennel La	14 C1	
Kenton Clo	15 E3	
Kibble Grn	16 E1	
Kiln La	14 C3	
Kimberley	16 E3	
Kingsmere Rd	14 B2	
Knightswood	16 D4	
Kyle Clo	14 C4	
Ladybank	16 D3	
Lakeside	15 E1	
Lambourn Gro	15 G3	
Larges Bridge Dri	15 E4	
Larges La	15 E3	
Latimer	16 D3	
Lauradale	14 C5	
Leafield Copse	15 H5	
Leaves Grn	16 F2	
Leicester	16 F2	
Lemington Gro	16 D1	
Leppington	16 D3	
Letcombe Sq	15 G5	
Lichfields	15 G3	
Lightwood	16 F2	
Lily Hill Dri	15 G3	
Lily Hill Rd	15 G3	
Lime Walk	15 E5	
Limerick Clo	14 C2	
Linden Hill Rd	14 B2	
Lingwood	16 E2	
Liscombe	16 D3	
Little Ringdale	15 G5	
Littledale Clo	15 F4	
Lochinver	16 D3	
London Rd	15 E3	
Long Hill Rd	15 H4	
Longmoors	14 A2	
Longshot La	14 A3	
Longwater Rd	16 E2	
Lovelace Rd	14 A5	
Lowbury	15 G5	
Ludlow	16 D3	
Lutterworth Clo	15 E1	
Lydbury	15 G4	
Lydney	16 D3	
Lyndhurst Clo	15 H4	
Lynwood Chase	15 E2	
Lytchett Minster Clo	15 G6	
Lytham	16 A2	
Madingley	16 D4	
Mainprize Rd	15 F2	
Makepiece Rd	14 D1	
Malham Fell	14 C5	
Manor Clo	14 C1	
Mansfield Cres	16 D2	
Manston Dri	16 D2	
Market St	14 D3	
Martins La	15 F4	
Meadow Way	14 B1	
Melrose	16 D4	
Membury Walk	15 G5	
Mendip Rd	15 G6	
Merlewood	15 E6	
Merryhill Rd	14 C1	
Micheldever Way	15 G6	
Milbanke Ct	14 B3	
Milbanke Way	14 B3	
Mill La	14 B5	
Milman Clo	15 H3	
Milton Clo	16 D2	
Minstead Clo	15 H4	
Moordale Av	14 A2	
Morden Clo	15 G5	
Mount La	15 E3	
Mount Pleasant	15 E4	
Naseby	16 D4	
Nelson Clo	15 G3	
Netherton	14 C5	
Nettlecombe	16 F1	
New Forest Ride	15 H6	
New Rd	15 F3	
Nightingale Cres	15 E6	
Nine Mile Ride	16 A4	
North Grn	15 F2	
Northampton Clo	15 F4	
Northcott	16 C4	
Nuthurst	15 F6	
Nutley	16 C4	
Oakdale	16 F2	
Oakengates	16 C4	
Oakwood Rd	15 G3	
Oareborough	15 F5	
Octavia	14 C4	
Old Bracknell Clo	14 C4	
Old Bracknell La	14 C4	
Old Bracknell La E.	14 D4	
Old Farm Dri	15 E1	
Oldbury	14 B4	
Oldstead	16 F1	
Ollerton	16 C3	
Opladen Way	16 F2	
Orion	16 C4	
Oxenhope	14 C5	
Pankhurst Dri	15 E6	
Park Rd	15 E3	
Parkland Dri	15 G2	
Peacock La	14 A4	
Pembroke	16 C3	
Pendine Pl	14 C6	
Pendlebury	16 C3	
Penwood Gdns	16 A2	
Peregrine Clo	14 D6	
Perry Oaks	15 G3	
Perry Way	15 G3	
Pewsey Vale	15 G6	
Pickering	14 C5	
Picket Post Clo	15 H4	
Ploughland	14 B2	
Pollardrow Av	14 B2	
Pond Moor Rd	14 D6	
Popham Clo	15 G6	
Portman Clo	14 C2	
Prescott	16 B3	
Priestwood Av	14 B2	
Priestwood Court	14 B2	
Priestwood Cres	14 C2	
Primrose Walk	15 E6	
Princess Sq	14 D3	
Priory La	15 E1	
Priory Walk	15 G5	
Qualitas	16 B4	
Quintilis	16 B4	

Radcliffe Way	14 A2
Radnor Rd	15 H4
Ralph's Ride	15 F4
Ramsbury Clo	16 A2
Ramslade Rd	15 E5
Ranelagh Dri	15 E4
Rapley Grn	16 E2
Rectory Clo	14 D5
Rectory La	14 C6
Rectory Row	14 D5
Redditch	16 F2
Redvers Rd	14 D6
Reeds Hill	16 D1
Rickman Clo	16 E1
Ringmead	16 A1
Ringwood	16 B3
Ripplesmere	15 F5
Rokeby Clo	15 E2
Rookswood	14 D1
Rosedale Gdns	14 C6
Rosset Clo	14 C5
Rounds Hill	14 A2
Rowley Clo	15 F4
Saffron Rd	14 C5
St Andrews	16 A2
St Anthonys Clo	14 C2
Salwey Clo	16 D2
Sandford Down	15 H6
Sandy La	14 D2
Sarum	16 A4
Savernake Way	15 G6
Scott Ter	15 G2
Segsbury Gro	15 G5
Setley Way	15 H4
Shaftesbury Clo	15 E6
Shelley Av	15 G3
Shepherds Hill	15 E2
Shepherds La	14 C1
Sherring Clo	15 E1
Sherwood Clo	15 H3
Silwood	16 A3
Simmonds Clo	14 A2
Skimped Hill La	14 C3
Smith Sq	15 F3
South Hill Rd	16 C2
South Lynn Cres	14 D6
South Rd	16 A3
Southern Ind Area	14 A4
Southwold	16 A3
Spencer Rd	14 B2
Spinis	16 A3
Spinner Grn	14 C6
Spring Meadow	15 F2
Stanley Walk	14 D3
Staplehurst	16 A3
Statham Ct	14 A2
Station Rd	14 D3
Staverton Clo	14 D1
Stokeford Clo	15 H6
Stoney Rd	14 B2
Stratfield	16 A4
Swaledale	16 C1
Swancote Grn	14 D6
Sweetwell Rd	14 A3
Sycamore Rise	15 F4
Sylvanus	16 B3
Tamworth	16 F3
Tawfield	16 A3
Tebbit Clo	15 E3
The Avenue	16 A4
The Cardinals	14 D5
The Close	15 E5
The Crescent	15 E5
The Croft	14 D1
The Green	14 D5
The Oaks	15 E3
The Paddock	15 E4
The Ridgeway	14 D4
The Ring	14 D3
Thornhill	15 F5
Threshfield	16 C1
Timline Grn	15 H3
Tippits Mead	14 A2
Toll Gdns	15 G4
Town Centre By-Pass	14 D2
Townsend Clo	15 G6
Trevelyan	16 A3
Trindle Down	14 C1
Troon Ct	16 A2
Trumbull Rd	14 C1
Turnberry	16 A2
Turnpike Rd	14 A2
Tytherton	15 E3

Uffington Dri	15 G5
Ullswater	16 A2
Underwood	16 A2
Upavon Gdns	15 G6
Upshire Gdns	15 G5
Vandyke	16 A2
Viking	16 A1
Vincent Rise	15 F4
Wagbullock Rise	16 E2
Walbury	15 G5
Waldron Hill	15 G2
Wallingford Clo	15 G5
Wantage Clo	15 F6
Wareham Rd	15 H6
Warfield Rd	15 E2
Waterham Rd	16 D2
Waterside Pk	14 A3
Wayland Clo	15 G5
Welbeck	16 A1
Wellington Dri	15 F6
West Gdn	15 F3
Westcombe Clo	15 F3
Westbrook Gdns	15 E2
Western Ind Area	14 B3
Western Rd	14 A2
Weycrofts	14 B1
Whatley Grn	16 D2
Wheatley	16 A1
Whistley Clo	15 F4
Whitton Rd	15 G4
Wickham Vale	16 A2
Wilberforce Way	15 E6
Wilders Clo	14 C1
Wildrings Rd	14 B6
Wildrings Sq	14 C5
Willoughby Rd	14 B4
Willow Dri	15 E2
Wilwood Rd	14 A2
Winchgrove Rd	14 C1
Windlebrook Grn	14 C2
Windlesham Rd	14 B2
Windmill Rd	14 B2
Winscombe	16 A1
Wittenham Rd	15 H2
Wokingham Rd	14 A2
Woodenhill	16 A4
Woodford Grn	15 G5
Woodland Cres	15 E1
Woodmere	15 F5
Woodridge Clo	14 D4
Woolhampton Way	15 F6
Wordsworth	16 A1
Wroxham	16 B1
Wychwood Av	15 H5
Wylam	16 B1
Yardley	16 B1

BRAY

Amerden La	17 D1
Beaufort Pl	17 C1
Bray Clo	17 B2
Bray Rd	17 A1
Braybank	17 C1
Brayfield Rd	17 C1
Church Dri	17 B1
Church La	17 B1
Ferry Rd	17 B1
Hanover Mead	17 B2
Hibbert Rd	17 A1
High St	17 B1
Monkey Island La	17 D2
Old Mill La	17 C1
River Gdns	17 C1
The Causeway	17 A1
The Terrace	17 B2
Upper Bray Rd	17 B2
Vicarage Dri	17 A1
Vicarage Walk	17 A1

BURGHFIELD COMMON

Abbey Park	17 A3
Abbots La	17 A3
Anstey Pl	17 C3
Ash La	17 B3
Auclum Clo	17 C4
Bannister Rd	17 A3
Barn Owl Way	17 C3
Birch Rd	17 B3
Blackbird Clo	17 C3
Blands Clo	17 A4
Bluebell Dri	17 A3
Boldrewood	17 B3
Bracken Way	17 B4
Brocas Rd	17 A4
Bunces La	17 B4
Burdock Clo	17 C3
Chervil Way	17 C3
Clay Hill Rd	17 A3
Dauntless Rd	17 C3
Field Clo	17 C3
Finch Way	17 C3
Firs End	17 B4
Fox Clo	17 B3
Garlands Clo	17 B4
Goodwood Clo	17 B4
Goring La	17 B4
Granby End	17 C3
Hawksworth Rd	17 D3
Headlands Ct	17 A4
Hermits Clo	17 C3
Hillside	17 D3
Hollybush La	17 A3
Holmdale	17 C3
Hunters Hill	17 B3
Jordans La	17 A3
Kennet Pl	17 C3
Kestrel Way	17 C3
Kirkwood Cres	17 A3
Lamsden Way	17 C3
Loves Clo	17 B3
Mans Hill	17 D3
Myrtle Clo	17 C3
Normoors Rd	17 A4
Oak Dri	17 B4
Oakdene	17 B3
Omers Rise	17 B3
Palmers La	17 B4
Pembroke Clo	17 D3
Pinchcut	17 B3
Pine Ridge Rd	17 B3
Ragdale	17 C3
Reading Rd	17 B4
Recreation Rd	17 B4
Robin Clo	17 C3
Russet Glade	17 C4
St Marys Way	17 C3
Saxby Clo	17 C3
School La	17 A3
Sorrel Clo	17 D3
Southwood Gdns	17 B3
Spring Wood La	17 C4
Stable Clo	17 B3
Sun Gdns	17 B4
Tarragon Way	17 C3
The Close	17 B3
Three Firs Way	17 A4
Thrush Clo	17 C3
Totterdown	17 A4
Valley Rd	17 C3
Warren Clo	17 B3
Wheeler Clo	17 C3
Woodlands Av	17 B3
Woodmans La	17 A3
Wren Clo	17 C3

COOKHAM

Abney Court Dri	19 H1
Alleyns La	18 D3
Bass Mead	19 F6
Bedwins La	18 B5
Berries Rd	19 H4
Bigfrith La	18 B5
Black Butts Cotts	19 H4
Bradcutts La	19 C2
Briar Glen	19 E5
Bridge Av	19 F5
Broom Hill	19 E5
Burnt Oak	19 F4
Cannondown Rd	19 E6
Causeway	19 G4
Choke La	18 B6
Church Rd	18 B5
Cookham Dean Bottom	18 C3
Coombe End	18 B5
Coxborrow Clo	19 F4
Danes Gdns	19 F5
Deans La	18 C3
Dedmere Rd	18 A1
East Paddock	18 A5
Elizabeth Clo	19 F4
Ferry La	19 H2
Fieldhouse La	18 A1
Fieldhouse Way	18 A1
First Av	18 A1
Firview Clo	18 A2
Fourth Av	18 A1
Gainsborough	19 F5
Gibralter La	18 C2
Gorse Rd	19 E5
Gossmore Clo	18 A2
Gossmore La	18 A2
Graham Rd	19 E5
Grange La	19 F3
Grange Rd	19 F4
Groves Way	19 F5
Grubwood La	18 B5
Halldore Rd	19 E4
Hardings Grn	18 C4
Hedsor Rd	19 H1
High Rd	19 E4
High St	19 H4
Hillcrest Av	19 E5
Hills La	18 D4
Hockett La	18 A5
Hollybush La	18 B5
Hyde Grn	18 A2
Inwood Clo	18 B5
Jeffries Ct	19 H1
Jobs La	18 C3
Kings La	18 C3
Lesters Rd	19 E5
Lightlands La	19 F6
Long La	18 C6
Lower Rd	19 E4
Lyndhurst Av	19 E5
Maidenhead Rd	19 F5
Marlow By-Pass	18 A3
Meadow Clo	18 A2
Mill La	19 H4
New Rd	19 E4
Newfield Gdns	18 A1
Orchard Hill	19 H2
Parkway	18 A1
Peace La	19 F5
Penling Clo	19 E5
Popes La	18 C3
Poundfield La	19 F4
Quarry Wood Rd	18 A3
Quarrydale Dri	18 A1
River Park Dri	18 A2
Riversdale	19 H3
Riverwood Av	18 B2
Riverwood Dri	18 A2
Roman Lea	19 F4
Rose Bank Clo	19 E4

Savill Way	18 A1
School La, Cookham	19 H4
School La, Cookham Dean	18 C4
Shergold Way	19 F5
South Wood Gdns	19 E6
Southwood Rd	19 E6
Spencers La	19 E4
Spring La	18 C6
Startins La	18 C3
Station Hill	19 F4
Station Rd	19 F4
Stone House La	18 D2
Strand La	19 F6
Stubbles La	18 B5
Sutton Clo	19 H4
Sutton Rd	19 H4
Switchback Rd	19 E6
Terrys La	19 E2
The Acre	18 A1
The Pound	19 G4
The Shaw	19 E5
Third Av	18 A1
Vicarage Clo	19 H4
Vivian Clo	19 F5
Wakelins End	19 F4
Walnut Tree Clo	19 H1
Walnut Way	19 H1
Warners Hill	18 D4
Weesons Hill	18 C4
Wessex Rd	19 H1
Westwood Grn	19 F5
Whiteladyes La	19 E4
Windmill Rd	19 E5
Winter Hill	18 B3
Winter Hill Rd	18 A5
Woodmoor End	19 H4
Worster Rd	19 E5

CROWTHORNE

Addiscombe Rd	21 G5
Albert Rd	21 F4
Alcot Clo	21 F5
Alderbrook Clo	20 D6
Aldworth Gdns	21 E3
Ardwell Clo	20 D4
Arenal Dri	21 G6
Ashdale Pk	20 B3
Badgers Sett	21 E4
Barracane Dri	21 F4
Barwell Clo	20 D5
Belmont Rd	21 F3
Benson Rd	20 D4
Birch Side	21 E3
Blake Clo	21 G5
Bracknell Rd	21 G4
Bramble Gate	21 E3
Bramley Ct	20 C5
Bramley Gro	20 C3
Brookers Corner	21 G4
Brookers Row	21 G3
Brunel Dri	21 G1
Butler Rd	21 F3
Byron Dri	21 F6
Cambridge Rd	21 G5
Celandine Clo	21 G3
Chaplains Hill	21 H5
Chaucer Rd	21 F6
Church Rd	21 F5
Church St	21 F4
Circle Hill Rd	21 G4
Coleridge Clo	21 G5
Copenhagen Walk	21 F5
Coppice Gdns	21 E4
Corsham Way	21 F4
Cricket Field Gro	21 H4
Dormer Clo	21 E4
Dukes Ride	20 C5
Dukes Wood	21 F4
Eagle Clo	21 E3
Edgecumbe Park Dri	21 E4
Edgedale Clo	21 F5
Edgewood Clo	21 E2
Elgar Av	21 F3
Ellis Rd	21 F3
Everest Rd	21 F3
Farm Clo	21 G2
Fern Clo	21 F2
Ferry Clo	21 E2
Fielding Gdns	21 F5
Fincham End Dri	20 D5
Forest Rd	21 G4
Fosse Way	21 F4
Frensham Rd	21 F3
Furzehill Cres	21 G5
Goldsmiths Way	21 F5
Gordon Rd	21 H6
Grange Av	21 F3
Grant Rd	21 G6
Greenfield Way	21 E2
Greenfinch Clo	21 E3
Greenside	21 E4
Greenwood Rd	21 E3
Greystoke Ct	21 F5
Grove Clo	20 C2
Hardy Grn	21 F5
Hatch Ride	21 E1
Heath Hill Rd	21 F4
Heath Ride	20 A3
Heather Way	21 E4
Heatherdene Av	20 C5
Heathermount Dri	21 E3
Heathermount Gdns	21 E3
Heathlands Rd	20 D1
High St	21 G5
Highway	21 E4
Hillary Dri	21 F3
Hinton Clo	21 F2
Hinton Dri	21 F2
Hollybush Ride	20 A5
Hollyhook Clo	21 E3
Holmbury Av	21 E2
Holme Clo	21 E2
Honey Hill	21 E1
Honeysuckle Clo	21 E2
Houston Way	20 C4
Keats Way	21 F3
Kentigern Dri	21 H4
Kings Rd	21 G5
Kingsley Clo	21 F6
Knowles Av	20 D4
Lake End Way	21 F5
Larkswood Dri	21 G4
Leacroft	21 F4
Leith Clo	21 F2
Leyside	21 E4
Linkway	21 E4
Little Fryth	20 B3
Llangar Gro	21 E4
Lower Broadmoor Rd	21 G5
Lower Wokingham Rd	20 B2
Lyneham Rd	21 F4
Lyon Rd	21 G2
Macadam Av	21 G2
Maple Clo	21 G2
Marigold Clo	21 E2
Merryman Dri	21 E3
Mulberry Clo	21 G5
Napier Clo	21 G5
Napier Rd	21 G5
New Rd	21 G4
New Wokingham Rd	21 E2
Nine Mile Ride	20 B2
Nugee Ct	21 G4
Oaklands La	21 E3
Old Pharmacy Clo	21 G5
Old Sawmill La	21 G3
Old Wokingham Rd	21 G1
Oleander Clo	21 E2
Otter Clo	21 E2
Palmer Clo	21 E1
Parkway	21 E4
Pensford Clo	21 F2
Pinefields Clo	21 F5
Pinehill Rd	21 G5
Pinewood Av	21 G3
Priors Wood	20 C5
Purcell Rd	21 F2
Range Rd	20 A2
Ravenswood Av	20 D5
Roman Ride	20 B4
Rowan Dri	21 G2
St Andrews Clo	20 D3
St Sabastians Clo	20 D1
Salamanca	20 C4
Sandhurst Rd	20 A1
Sandhurst Rd	21 F6
School Rd	21 H5
Seymour Ct	20 C5
Shaw Pk	21 F6
Shepherds Way	20 C5
Soldiers Rise	20 B2
South Meadow	21 H6
Squirrels Drey	21 E4
Sweetbriar	21 E2
Sydney Clo	21 G2
Telford Av	21 G2
The Avenue	21 F3
The Bracken	21 E2
The Brambles	20 C3
The Chase	21 E3
The Conifers	21 E2
The Devils Highway	20 D4
The Paddock	21 E3
The Rise	21 E4
The Terrace	21 H4
Thornbury Clo	21 F4
Thorne Clo	21 E2
Timberley Pl	20 C4
Towers Dri	21 F5
Upper Broadmoor Rd	21 G4
Walmer Clo	21 G4
Waterloo Rd	21 E5
Wellesley Dri	20 C4
Wellington Rd	21 G5
Wellingtonia Av	20 A5
Wentworth Clo	21 E2
Westbury Clo	21 F4
White City	21 H4
Wiltshire Av	21 F3
Wokingham Rd	20 C5
Wood End	20 D5

DATCHET

Adelaide Rd	22 A2
Beaulieu Clo	22 B2
Buccleuch Rd	22 B2
Cobb Clo	22 D2
Datchet Pl	22 C2
Ditton Rd	22 D2
Dutch Elm Av	22 A2
Elmcroft	22 C2
Eton Clo	22 A1
Eton Rd	22 A1
Fairfield Av	22 C1
Fairfield Clo	22 D1
Gables Clo	22 A1
Green La	22 C2
Hall Ct	22 B1
High St	22 B2
Holmlea Rd	22 D2
Holmlea Walk	22 D2
Horton Rd	22 B2
Lawn Clo	22 C1
Leigh Pk	22 C1
Lime Av	22 A2
Linchfield Rd	22 C2
Link Rd	22 C2
London Rd	22 B1
Majors Farm Rd	22 D1
Manor House La	22 B2
Marshfield	22 C2
Mill Pl	22 D2
Montague Rd	22 B2
Montrose Av	22 C1
Montrose Way	22 D2
New Rd	22 D2
Penn Rd	22 D2
Percy Pl	22 C2
Prince Alberts Walk	22 A1
Priory Way	22 B1
Queen Victorias Walk	22 A2
Queens Mead	22 B2
Queens Rd	22 B2
Riding Court Rd	22 C1
Ruscombe Gdns	22 B1
Saffron Clo	22 C2
Slough Rd	22 B1
Southlea Rd	22 B2
Talbot Pl	22 D2
The Avenue	22 B2
The Drive	22 C2
The Green	22 B2
The Paddock	22 B2
Whites La	22 B1
Windsor Rd	22 A1

HUNGERFORD

Atherton Cres	23 B4
Atherton Rd	23 B4
Bearwater	23 C1
Bourne Vale	23 A4
Bridge St	23 C2
Bulpit La	23 C5
Canal Walk	23 C2
Chantry Mead	23 A4
Chapel Ct	23 C1
Charnham La	23 C1
Charnham Park	23 B1
Charnham St	23 B1
Cherry Gro	23 B3
Chestnut Walk	23 C6
Chilton Way	23 B3
Church Croft	23 B2
Church St	23 B3
Church Way	23 B3
Clarks Gdns	23 C4
Coldharbour Rd	23 C5
Coombe Vw	23 C6
Croft La	23 C2
Croft Rd	23 B3
Crown Mews	23 C3
De Montfort Gro	23 B5
Eddington Hill	23 D1
Everland Rd	23 C2
Fairfields	23 C4
Fairview Rd	23 C4
Faulkner Sq	23 C1
Freemans Clo	23 A3
High St	23 C4
Hillside Rd	23 C5
Homefield Way	23 B4
Honeyfields	23 B3
Inkpen Rd	23 D6
Ligueil Clo	23 C6
Lancaster Clo	23 B5
Lancaster Sq	23 B5
Macklin Clo	23 C3
Marsh La	23 A3
Moores Pl	23 B3
Morley Pl	23 C4
North Vw	23 C3
Orchard Park Clo	23 C4
Oxford St	23 D1
Park St	23 C3
Parkway	23 C5
Parsonage La	23 B3
Pound Piece	23 A3
Priory Av	23 C6
Priory Clo	23 C5
Priory Rd	23 B4
Priory Ter	23 C4
Prospect Rd	23 B4
Regent Clo	23 B4
St Lawrence Sq	23 B3
Salisbury Rd	23 A6
Sanden Clo	23 B4
Sarum Way	23 B4

Shalbourne Clo	23 A3
Smitham Bridge Rd	23 A3
Somerset Clo	23 A3
South Vw	23 C4
Station Rd	23 C3
Strongrove Hill	23 A1
Tarrants Hill	23 C4
The Croft	23 B2
Uplands	23 B4
Wessex Clo	23 A3
Westbrook Clo	23 A4
York Rd	23 B5

LAMBOURN

Aintree	22 C4
Atherton Pl	22 B3
Baydon Rd	22 A4
Big La	22 B3
Blind La	22 C4
Bockhampton Rd	22 C4
Child St	22 B3
Close End	22 B4
Crowle Rd	22 B4
Derby Clo	22 B4
Edwards Hill	22 B4
Folly Rd	22 A3
Foxbury	22 C4
Goose Grn	22 C3
Greenway	22 B4
Gwyns Piece	22 C3
Harris Clo	22 C3
High St	22 B4
Honey Hill	22 C3
Hungerford Hill	22 B4
Lambourn Pl	22 B3
Lynch La	22 B3
Mill La	22 C3
Millfield	22 C3
Montague St	22 C3
Northfields	22 C3
Newbury Rd	22 C4
Newbury St	22 C4
Oxford St	22 B3
Parsonage Pl	22 B3
Parsonage Rd	22 B3
Pegasus Ct	22 C3
Rockfell Rd	22 B4
St Michaels Clo	22 B3
Sheepdrove Rd	22 C3
Station Rd	22 C4
The Broadway	22 C3
The Classics	22 C4
The Granthams	22 C3
The Park	22 B3
Tubbs Farm Clo	22 C4
Upper Lambourn Rd	22 A3
Walkers La	22 C3
Wantage Rd	22 C3
Woodbury	22 C4

MAIDENHEAD

Addison Ct	25 H2
Albert St	25 F5
Aldebury Rd	25 F1
Aldwick Dri	24 D5
Alexandra Rd	24 D3
Alienby Rd	24 B4
All Saints Av	24 D3
Altwood Bailey	24 C6
Altwood Clo	24 B6
Altwood Dri	24 B6
Altwood Rd	24 A6
Alwyn Rd	24 B3

Anne Clo	25 E1
Archer Clo	24 D3
Arlington Clo	24 A3
Arundel Clo	24 A3
Ashcroft Rd	24 C3
Ashley Park	25 H1
Ashton Pl	24 B5
Aston Clo	25 H5
Athlone Clo	25 E2
Atkinsons All	25 F3
Auckland Clo	25 H3
Audley Dri	24 C5
Australia Av	25 F3
Autumn Walk	24 A5
Avenue Rd	25 H6
Avondale	24 C2
Bad Godesberg Way	25 F4
Badminton Rd	24 C5
Bailey Clo	25 F4
Bakers La	24 A3
Balmoral	24 C2
Bannard Rd	24 A6
Bargeman Rd	25 E6
Barn Clo	25 F1
Bath Rd	24 A5
Beechwood Dri	24 B5
Bell St	25 F5
Belmont Cres	24 D3
Belmont Dri	24 D3
Belmont Park Av	24 D2
Belmont Park Rd	24 D3
Belmont Rd	24 D3
Belmont Vale	24 D3
Berkeley Clo	24 A3
Beverley Gdns	24 B2
Birdwood Rd	24 B4
Bishop Ct	24 D5
Bix La	24 A2
Blackmore La	25 G3
Blakeney Ct	25 F2
Blenheim Rd	24 B3
Blomfield Rd	24 B6
Boyn Alley Rd	24 D6
Boyn Gro	24 C5
Boyn Hill Av	24 D5
Boyn Hill Clo	24 D5
Boyn Hill Rd	24 D6
Boyndon Rd	25 E4
Brampton Ct	25 H3
Bray Rd	25 H5
Braywick Rd	25 F5
Briar Dene	24 C2
Bridge Av	25 G4
Bridge Rd	25 G4
Bridge St	25 G4
Bridle Clo	25 C2
Bridle Rd	25 E2
Broadway	25 F5
Brock La	25 F4
Brompton Dri	24 C1
Brookdene Clo	25 F1
Brunel Clo	25 E6
Brunel Rd	25 E6
Calder Clo	25 E2
Calder Ct	25 E2
Camden Rd	24 D2
Camley Gdns	24 A3
Camley Park Dri	24 A3
Camperdown	25 H2
Cannock Clo	25 H5
Cannon Court Rd	24 D1
Cannon La	24 A6
Cannon Rd	24 D1
Carisbrook Clo	24 C5
Castle Ct	25 E4
Castle Dri	25 E4
Castle Hill	25 E4
Castle Hill Ter	25 E4
Castle Ms	25 E4
Cedars Rd	25 G4
Chalgrove Clo	25 H5
Challow Ct	25 E2
Chatsworth Clo	24 C6
Chauntry Rd	25 H5
Cheniston Grn	24 A4
Cherington Gate	24 B2
Cherwell Clo	25 G3
Chestnut Clo	25 H2
Cheviot Clo	25 H5
Chiltern Rd	25 H5
Church Rd	25 H6
Clappers Meadow	25 H2
Clare Rd	24 D5
Clarefield Clo	24 B2

Clarefield Dri	24 B2
Clarefield Rd	24 C2
Cleveland Clo	25 H5
Cliveden Mead	25 H1
Clivemont Rd	25 F2
College Av	25 E4
College Glen	24 D4
College Rise	24 D4
College Road	25 E3
Collier Clo	25 F2
Coln Clo	25 F3
Connaught Clo	25 E2
Cookham Rd	25 F1
Cordwallis Rd	25 E3
Cordwallis St	25 E3
Corfe Pl	24 C4
Cornwall Clo	25 E1
Cotswold Clo	25 H5
Courtfield Dri	24 C5
Courthouse Rd	24 C4
Courtlands	25 F5
Cranbrook Dri	24 B2
Craufurd Rise	25 E3
Creden Clo	24 D3
Crescent Dri	25 E4
Crompton Dri	24 A3
Cromwell Rd	24 D4
Crown La	25 G4
Croxley Rise	24 D5
Culham Rd	25 E1
Deerswood	25 H2
Denham Clo	24 C5
Denmark St	25 F3
Depot Rd	25 F5
Derwent Dri	24 D3
Desborough Cres	24 D6
Dhoon Rise	25 F5
Dorchester Clo	24 C2
East Rd	25 F4
Edinburgh Rd	25 E2
Edith Rd	24 B4
Ellington Pk	25 E2
Elm Gro	25 E4
Elton Dri	25 E3
Evenlode	25 F3
Fair Acre	24 D5
Fairford Rd	25 F3
Fane Way	24 D6
Farm Clo	24 A4
Farm Rd	24 A4
Farmers Way	24 A6
Fawley Ct	25 E1
Fernley Ct	25 E2
Fielding Rd	24 C4
Finch Ct	24 D6
Florence Av	25 G3
Fontwell Clo	24 A3
Forlease Ct	25 G5
Forlease Dri	25 G5
Forlease Rd	25 G4
Fotherby Ct	25 G5
Frascati Way	25 F4
Fullbrook Clo	25 G3
Furze Platt Rd	24 A1
Furze Rd	25 E2
Gables Clo	25 H3
Gardner Rd	25 E1
Garthlands	24 D1
Glebe Rd	25 H6
Gloucester Rd	25 E1
Gordon Rd	24 D4
Grafton Clo	25 E1
Graham Clo	24 C6
Grassy La	25 E4
Grays Alley	24 A4
Great Hill Cres	24 C6
Green Clo	25 F2
Green La	25 G5
Greenfields	25 G5
Greenways Dri	24 B3
Grenfell Av	25 F5
Grenfell Pl	25 F5
Grenfell Rd	25 E4
Griffin Clo	25 E6
Gringer Hill	25 E2
Grove Rd	25 G4
Gwendale	24 C2
Haddon Rd	24 C6
Halifax Clo	24 B3
Halifax Rd	24 B3
Halifax Way	24 B3

Hamilton Pk	24 B5
Hampden Rd	24 B3
Hardwick Clo	24 A3
Hare Shots	25 E6
Harefield Rd	24 B4
Hargrave Rd	25 E3
Harrow Clo	25 F2
Harrow La	25 E2
Hatfield Clo	24 C5
Havelock Cres	24 C4
Havelock Rd	24 C4
Hazel Clo	25 G3
Headington Clo	24 A4
Headington Rd	24 A3
Heathlands Dri	24 A5
Helmsdale	24 C2
Henley Rd	24 A5
Hever Clo	24 C5
High St	25 F4
High Town Rd	25 E5
Highfield Rd	24 C3
Highgrove Pk	25 E3
Highway Av	24 B4
Highway Rd	24 B5
Hillside	25 E6
Hobbis Clo	24 B5
Holman Leaze	25 G3
Holmwood Clo	24 B6
Horseguards Dri	25 H4
Howarth Rd	25 G5
Hughendon Clo	24 C5
Ilchester Clo	24 D6
In The Ray	25 H3
Juniper Dri	25 H3
Keble La	25 E3
Kennedy Clo	24 C5
Kennet Rd	25 F3
Kent Way	25 E2
Kenwood Clo	24 B4
Keys La	25 F5
Kidwells Clo	25 F3
King St	25 F4
King St	25 F5
Kings Dri	25 E5
Kings Gro	25 E5
Kingswood Ct	25 G6
Knowlsey Clo	24 B2
Laburnham Rd	25 E5
Laggan Rd	25 F2
Laggan Sq	25 F2
Lakeside	25 H2
Lancaster Rd	24 B3
Lancastria Mews	25 E4
Langdale Clo	25 G5
Langton Clo	24 D3
Lantern Walk	25 H4
Larchfield Rd	25 E6
Lassell Gdns	25 H4
Lee La	24 A2
Lees Clo	24 B6
Lees Gdns	24 B6
Leighton Gdns	25 H2
Lexington Av	24 D6
Lime Walk	24 A3
Lincoln Rd	24 B3
Linden Av	24 D2
Lissett Rd	25 G5
Loddon Dri	24 D3
Longleat Gdns	24 D5
Lonsdale Clo	25 G2
Lower Boyndon Rd	25 E5
Ludlow Rd	25 E5
Lutmans La	25 F1
Lyneham Gdns	24 C2
Lynton Grn	25 E4
Mallow Pk	24 C2
Malvern Rd	25 D2
Maple Clo	24 D6
Market St	25 F4
Marlborough Clo	24 B5
Marlborough Rd	24 B5
Marlow Rd	25 F4
Martin Clo	25 F3
Meadway	25 G4
Medallion Pl	25 H4
Melton Ct	25 F5
Michael Clo	24 D6
Moffey Hill	25 E1
Moneycrower Dri	25 E4
Montrose Dri	24 A5
Moor La	25 F2

Moorbridge Rd	25 G4	Silver Clo	24 A6	Bagnols Way	26 B6
Moorfield Ter	25 H3	Silvertrees Dri	24 B6	Balfour Cres	28 A5
Moorlands Dri	24 A3	Simpson Clo	25 H3	Barn Cres	28 B4
Moorside Clo	25 F2	South Rd	25 F5	Bartholomew St	26 D6
Mossy Vale	25 E2	Spencers Clo	24 D3	Bartlemy Clo	28 C2
Muddy La	24 C5	Spencers Rd	24 D3	Bartlemy Rd	28 C2
Murrin Rd	24 C3	Sperling Rd	25 F2	Bath Rd	26 A3
		Spring Clo	25 F1	Battery End	28 B4
Napier Rd	24 B5	Stafferton Way	25 G5	Battle Clo	26 B3
Newbury Dri	25 H5	Stamford Rd	24 D5	Battle Rd	28 A4
Nicholsons La	25 F4	Station App	25 F5	Bedford Clo	28 A5
Norden Rd	24 D6	Stonefield Park	24 D4	Bell Hill	28 A6
Norfolk Park Cotts	25 F3	Summerford Ct	25 H3	Bell Holt	28 A5
Norfolk Rd	25 E3	Summerleaze Rd	25 G2	Belvedere Dri	28 D2
North Dean	25 F2	Sun La	25 F4	Benett Clo	26 C3
North Field Rd	25 F2	Sunderland Rd	24 B3	Berkeley Rd	26 C5
North Green	25 F2	Sutton Clo	24 D5	Birchwood Rd	27 G4
North Rd	25 E4	Switchback Clo	24 D1	Black Bear La	26 D5
North Star La	24 D5	Switchback Rd Nth	25 E1	Bledlow Clo	28 A5
North Town Clo	25 G2	Switchback Rd Sth	24 D1	Blenheim Rd	26 C5
North Town Mead	25 G2	Sylvester Rd	25 E1	Bone La	27 E5
North Town Moor	25 F2			Bonemill La	28 B1
North Town Rd	25 F2	Tachbrook Clo	25 G3	Bostock Rd	29 E2
		Talbots Dri	24 B5	Boundary Rd	27 E5
Oaken Gro	24 C2	Tavistock Clo	24 A3	Braunfels Walk	26 B6
Oldacres	25 H4	Thames Cres	25 H1	Bridge St	26 D5
Oldershaw Mews	24 C3	Thatchers Dri	24 A6	Bronte Rise	29 E2
Oldfield Rd	25 H5	The Crescent	25 E4	Brookway	27 H6
Orchard Gro	24 D4	The Croft	24 D6	Bruan Rd	28 D3
Osney Rd	25 E1	The Farthingales	25 H4	Brummell Rd	26 B3
Ostlergate	24 C2	The Paddock	24 C1	Buckingham Rd	28 C1
		The Pagoda	25 H2	Bunkers Hill	28 A5
Park St	25 G5	The Ridings	24 A4	Burchell Rd	26 B3
Parkside	24 D2	Thicket Grn	24 A4	Burys Bank Rd	29 G3
Partridge Mead	25 F1	Timbers Walk	24 B6	Butson Clo	26 B5
Pearce Clo	25 F2	Tollgate	24 A5	Byron Clo	28 D3
Pearce Rd	25 F2	Truro Clo	24 A4		
Penshurst Rd	24 D6	Tuffins Clo	24 B6	Cansfield End	26 C5
Penyston Rd	24 C4	Twynham Rd	24 B4	Carey Clo	28 B4
Pine Clo	24 C4			Carnegie Rd	26 D6
Pinkneys Dri	24 A3	Underhill Clo	25 E5	Castle Gro	26 C3
Pinkneys Rd	24 B2			Castle La	26 C2
Poplars Gro	25 H1	Vicarage Rd	25 F3	Catherine St	29 E1
Portlock Rd	24 C4			Caunter Rd	26 B3
Powney Rd	24 D4	Waldeck Rd	25 H4	Cavalier Clo	27 F3
Prince Andrews Clo	25 H2	Wavell Rd	24 C5	Cedar Mt	28 D3
Prince Andrews Rd	25 H2	Wayside Ms	25 F3	Chalford Rd	28 B5
Princess St	25 F5	Webster Clo	24 B6	Chandos Rd	28 D3
Providence Pl	25 G4	Webster Ct	24 D5	Charles St	28 B4
		Welbeck Rd	24 D6	Charlton Pl	27 E4
Queen St	25 F5	Wellhouse Rd	25 E1	Charnwood Clo	26 C3
Queensway	25 E2	Wellington Rd	24 D4	Charter Rd	28 D3
		Wentworth Cres	24 C5	Chaucer Clo	26 B3
Ray Dri	25 H4	West Dean	25 F3	Chaucer Cres	26 B3
Ray Hill Rd W	25 F3	West Rd	25 E4	Cheap St	26 D5
Ray Lea Clo	25 H3	West St	25 F4	Cheriton Clo	29 E2
Ray Lea Rd	25 H2	Westborough Rd	24 D5	Cherry Clo	26 D3
Ray Mill Rd	25 G2	Westfield Rd	24 C4	Chester Clo	29 G2
Ray Mill Rd E	25 H2	Westmead	25 F1	Chesterfield Rd	28 D1
Ray Park Av	25 H2	Westmorland Rd	24 D4	Chestnut Cres	26 D4
Ray Park La	25 H3	White Hart Rd	25 F4	Cheviot Clo	28 A5
Ray Park Rd	25 H3	White Rock	25 H2	Chiltern Clo	28 A5
Ray St	25 H4	Whurley Way	25 E1	Christie Heights	29 E3
Raymond Rd	24 D4	Winbury Ct	25 E4	Church La	26 A4
Redriff Clo	24 D5	Windrush Way	25 G3	Church Rd	27 E3
Reform Rd	25 H4	Winter Hill Rd	24 A1	Clarendon Gdns	26 D4
Reid Av	25 E6	Woodcote	25 E5	Clay Hill	27 G3
Risborough Rd	25 F3	Woodfield Dri	24 B5	Cleveland Gro	26 C5
Riseley Rd	24 D4	Woodhurst Rd	25 H2	Clifton Rd	26 B6
Rixman Clo	24 D6	Woodstock Clo	25 F2	Collins Clo	27 F4
Roseleigh Clo	24 B4	Wootton Way	24 C5	Conifer Crest	28 A5
Ross Rd	25 E6			Connaught Rd	27 E5
Rushington Av	25 F6	Yew Tree Clo	25 E3	Cope Hall La	28 A3
Russell Ct	25 F4	York Rd	25 G5	Coppice Clo	29 E2
Rutland Pl	24 D5			Coster Clo	26 B3
Rutland Rd	24 D5	**NEWBURY**		Courtlands Rd	29 E2
				Cowslade Rd	26 B3
Sadlers Mews	25 H4	Abbey Rd	29 E2	Coxeter Rd	26 B3
St Cloud Way	25 G4	Abbots Rd	29 E1	Cranford Pl	26 C5
St Ives Rd	25 G4	Aintree Clo	29 F2	Craven Rd	26 B6
St Lukes Rd	25 F4	Albert Rd	26 D4	Crawford Rd	26 C5
St Margarets Rd	24 B4	Almond Av	26 D3	Creswell Rd	27 G4
St Marks Cres	24 B4	Amberley Clo	26 C4	Croft La	26 C4
St Marks Rd	24 C4	Ampere Rd	27 E4	Croft Rd	28 D2
St Peters Rd	24 D1	Andover Rd	28 B6	Cromwell Pl	26 D5
Salters Clo	25 H4	Angel Ct	26 D4	Cromwell Rd	27 F3
Salters Rd	25 H4	Apex Rd	27 F5	Cromwell Ter	26 A3
Sandisplatt Rd	24 A5	Appletree Clo	28 C3	Culver Rd	28 C3
Sandringham Rd	25 F1	Argyll Rd	28 D1	Curling Way	27 F4
School La	25 F2	Arnhem Rd	27 E5	Cyril Vokins Rd	27 H6
Sheephouse Rd	25 H2	Arthur Rd	28 C1		
Sherwood Dri	24 B5	Ascot Clo	29 F2	Dalby Cres	29 F2
Shifford Cres	25 E1	Ashton Rd	27 E5	De Montfort Rd	26 B3
Shirley Rd	24 C6	Ashwood Dri	27 G4	Deadmans La	29 E4
Shoppenhangers Rd	25 E6	Audley Clo	27 G3	Dene Way	26 D2
Silco Dri	25 F5	Austen Gdns	29 F2	Denmark Rd	27 E5

Derby Rd	28 D1		
Dickens Walk	29 E2		
Digby Rd	26 C3		
Dolman Rd	26 D3		
Donnington Pk	26 C2		
Donnington Sq	26 C3		
Dormer Clo	28 B4		
Doveton Way	27 E4		
Dysons Clo	26 C5		
Edgecombe La	27 F3		
Eeklo Pl	29 E1		
Elizabeth Av	28 B4		
Enborne Gro	28 C1		
Enborne Pl	28 C1		
Enborne Rd	28 A1		
Enborne St	28 A5		
Epsom Cres	29 F2		
Erleigh Dene	28 D2		
Erving Walk	29 E2		
Essex St	28 A4		
Express Way	27 H6		
Falkland Dri	28 D2		
Falkland Garth	28 B4		
Falkland Rd	28 B4		
Faraday Rd	27 E5		
Fennel Clo	27 G3		
Fieldridge	27 F3		
Fifth Rd	28 C2		
Fir Tree La	27 H4		
First Av	27 F4		
Fleming Rd	27 E4		
Fontwell Rd	29 F1		
Friars Rd	29 E2		
Garden Close La	28 B5		
Garford Cres	28 B3		
Gaywood Dri	27 G4		
Gilray Clo	28 A5		
Glendale Av	28 A5		
Gloucester Rd	26 B6		
Goldwell Dri	26 C4		
Goodwin Walk	28 A4		
Goodwood Way	29 F1		
Gordon Rd	27 E6		
Gorselands	28 B5		
Green La	26 B6		
Greenham Mill	27 E5		
Greenham Rd	29 E1		
Greenlands Rd	29 E2		
Greyberry Copse	29 G3		
Grove Rd	26 B3		
Groveland Rd	26 B3		
Gwyn Clo	28 C2		
Hambridge La	27 G5		
Hambridge Rd	27 F6		
Hamilton Ct	29 E2		
Hampton Rd	28 D1		
Harvest Grn	28 C1		
Hawthorn Rd	26 D4		
Hedgeway	27 F4		
Henshaw Cres	28 B2		
Hereward Clo	26 C3		
Highfield Rd	28 D2		
Highwood Clo	27 F2		
Hill Clo	28 B4		
Hill Rd	26 B4		
Holbourne Clo	28 A5		
Homemead Clo	28 B3		
Hopwood Clo	27 G4		
Howard Rd	29 E1		
Hutton Clo	27 E4		
Jesmond Dene	26 C4		
Jubilee Rd	27 E6		
Kelvin Rd	27 E4		
Kempton Clo	29 F2		
Kendrick Rd	28 B5		
Kennedy Clo	28 B4		
Kennet Rd	26 C5		
Kennet Side	27 F5		
Kersey Cres	26 B3		
Kiln Rd	27 E3		
Kimbers Dri	26 A3		
Kingfisher Ct	27 G5		
Kings Rd	27 E5		
Kings Rd W	26 D6		
Kingsbridge Rd	28 B1		
Kingsland Grange	28 C3		
Kingsley Clo	27 E2		
Laburnham Gro	26 D4		
Ladwell Clo	28 B6		

Lambourn Rd	26 A2	Racecourse Rd	29 E1	Westbourne Ter	26 D4
Lamp Acres	27 F2	Railway Rd	27 E6	Western Av	26 C4
Leslie Southern Ct	27 E4	Rectory Clo	28 D1	Western End	26 B6
Lewendon Rd	26 B3	Redfield Ct	27 G4	Westfield Way	26 B6
Leys Gdns	26 C4	Regnum Dri	27 E3	Westgate Rd	28 C1
Lime Clo	27 G4	Rememberance Rd	28 C1	Westlands Rd	29 E2
Linden Clo	26 C4	Ring Rd	27 E5	Westmead Dri	28 D2
Lingfield Rd	29 F2	Robertson Clo	29 E3	Westwood Rd	29 F2
Lington Clo	27 F3	Robins Clo	28 D3	Wharf Rd	26 D5
Link Rd	28 D1	Rockingham Rd	26 C6	Wharf St	26 D5
Lipscombe Clo	26 B5	Roebuts Clo	28 D2	Wheatlands La	28 A4
Lisle Clo	26 D3	Rokeby Clo	29 E3	Willow Clo	28 D1
Livingstone Rd	27 E6	Round End	28 B5	Willowmead Clo	28 A6
London Rd	27 E4	Rowan Dri	26 D3	Wilmot Walk	28 A4
Long La	27 F2	Rupert Rd	28 D3	Winchcombe Rd	26 D5
Longacre	28 B2	Russell Rd	26 B5	Windsor Rd	29 F2
Love La	26 C2			Winterton Dri	26 A3
Lower Way	27 H4	Saffron Clo	26 C5	Woodridge	28 C3
Lynton Ct	26 D4	St Davids Rd	26 C6	Woodside	28 B3
		St Donats Pl	29 E1	Wyndham Rd	27 G3
Majendie Clo	26 A3	St Georges Av	26 C6		
Manor La	27 H4	St Johns Rd	28 D1	York Clo	29 E1
Manor Pl	26 A3	St Marys Rd	27 E4	York Rd	29 E1
Mansell Dri	28 A5	St Michaels Rd	26 C6		
Mansion House St	26 D5	St Nicholas Rd	26 C6		
Maple Cres	26 D3	Salcombe Rd	28 C1	**OLD WINDSOR**	
Marconi Rd	27 E4	Sandleford Link	29 E1		
Market Pl	26 D5	Sandleford Rise	28 D3	Albany Rd	45 A3
Market St	26 D6	Sandown Way	29 F1	Albert Rd	45 A2
Marsh La	26 D5	Sayers Clo	29 F3	Ashbrook Rd	45 B5
Marshalls Ct	26 A3	Sedgefield Rd	29 F2	Aylesworth Spur	45 B5
Meadow Rd	28 D2	Shaw Farm Rd	27 E2		
Meldrum Clo	28 A5	Shaw Hill	27 E3	Burfield Rd	45 A4
Meyrick Dri	28 A5	Shaw Rd	27 E4		
Middle Clo	28 B3	Sherrardmead	27 E3	Cell Farm Av	45 B3
Mill La	27 E5	Shop La	26 C2	Church Rd	45 B3
Monks La	28 C4	Sidestrand Rd	28 C3	Clayhall La	45 A3
Monkswood Clo	28 B3	Skipings Clo	28 A4	Coppice Dri	45 D5
Montague Ter	28 D2	Skyllings	27 F4	Cornwall Rd	45 A4
Montgomery Rd	28 C2	Smallridge	28 A6	Crimp Hill	45 A4
Moor La	26 B4	Sorrel Clo	27 G3	Crimp Hill Rd	45 A5
Mount Clo	29 E2	Speen Hill Clo	26 B4		
Mountbatten Clo	27 F3	Speen La	26 A3	Datchet Rd	45 A2
		Speen Lodge Ct	26 B4		
New Rd	29 F1	Speen Pl	26 B4	Fairfield App	45 D4
Newbold Rd	26 B3	Spencer Rd	28 B5	Fairfield Rd	45 D4
Newbury Business Pk	27 F4	Springfield La	29 E3	Follet Clo	45 B4
Newport Clo	27 E4	Stable Ct	27 E2	Friary Rd	45 C5
Newport Rd	27 E4	Stanley Rd	27 E6		
Newtown Rd	28 D1	Stapleton Clo	28 A4	Garston La	45 D5
Normay Rise	28 A5	Station Rd, Newbury	26 D6	Glebe Rd	45 B3
Northbrook St	26 D4	Station Rd, Speen	26 A3	Gregory Dri	45 B4
Northcroft La	26 D5	Stoney La	27 G3	Grove Clo	45 C5
Northern Av	26 D2	Strawberry Hill	26 C4		
Northway	27 E6	Stuart Rd	28 A4	Harwood Gdns	45 B5
Northwood Dri	27 F3	Sunley Clo	28 B5		
Norton Clo	28 A5	Sutherlands	28 C4	Keppel Spur	45 B5
		Sutton Rd	26 B3	King Johns Clo	45 C4
Oaken Gro	28 B2	Sycamore Rise	27 F3	Kingsbury Dri	45 B4
Oakley Rd	27 G4	Sylvester Clo	26 B3	Kingswood Creek	45 D3
Oddfellows Rd	26 D5				
Old Newtown Rd	28 D1	The Baxendales	29 F1	Lyndwood Dri	45 A4
Orchard Clo	27 F3	The Brambles	28 C3		
Orchardene	27 E4	The Broadway	26 D4	Magna Carta La	45 D6
Orts Rd	27 E5	The Drive	28 C2	Malt House Clo	45 C5
Owen Rd	27 E2	The Folly	29 E1	Meadow Clo	45 B4
Oxford Rd	26 C3	The Gabriels	28 B5	Meadow Way	45 B4
Oxford St	26 D4	The Glade	28 C3	Millers La	45 A4
		The Hollies	28 B5	Mills Spur	45 B5
Paddock Rd	28 D2	The Laurels	28 C3		
Park End	26 D4	The Maltings	26 D5	Newton Ct	45 A4
Park Rd	26 D4	The Marlows	28 D3	Newton La	45 B4
Park St	26 D4	The Nightingales	29 E2	Nursery Way	45 D4
Park Way	26 D4	The Paddock	27 F6		
Parsons Clo	26 B6	The Sydings	26 A3	Old Ferry Dri	45 C4
Pear Tree La	27 F2	Three Acre Rd	28 C3	Orchard Rd	45 B4
Pelican La	26 D4	Trout Walk	27 F3	Ouseley Rd	45 C5
Pembroke Rd	26 D5	Tudor Rd	29 E1		
Penn Rd	26 B3	Turnpike Rd	27 G3	Park Av	45 D3
Penrose Clo	26 C3	Tydehams	28 C3	Pelling Hill	45 B5
Pigeon Farm Rd	29 G3			Pollards Clo	45 B3
Pike St	27 F4	Valley Rd	28 B3	Priest Hill	45 C6
Pinchington La	29 E3	Villiers Way	28 A4		
Pine Ridge	27 G3	Vine Ct	29 E1	Queens Clo	45 B3
Plumpton Rd	29 F2				
Pond Clo	28 B3	Waldergrave Pl	26 D5	Ricardo Rd	45 B4
Poplar Pl	26 D3	Walton Way	27 F3	Riverside	45 C5
Porchester Rd	28 D1	Warren Rd	28 B5	Robin Willis Way	45 A4
Porter End	29 F2	Warwick Dri	29 F1		
Posting House Ms	26 C4	Water La	29 G3	St Andrews Clo	45 A4
Pound La	26 A3	Well Meadow	27 E3	St Lukes Rd	45 A4
Pound St	26 C6	Wendan Rd	28 D2	St Peters Clo	45 B3
Priory Rd	28 D2	Wentworth Ct	29 E1	Saxon Way	45 B4
Prospect Pl	28 D1	West Mills	26 D5	Southlea Rd	45 A2
Puffers Way	26 B6	West Mills Yd	26 D5	Straight Rd	45 A3
Pyle Hill	29 F2	West St	26 D5		
Queens Rd	27 E6				

The Avenue	45 B3
The Crofters	45 A4
The Drive	45 D3
The Embankment	45 C5
The Friary	45 C4
The Grange	45 B3
Tudor La	45 C5
Walpole Rd	45 B5
Warrington Spur	45 B5
Wharf Rd	45 C5
William Ellis Clo	45 A3
Windsor Rd	45 C6

PANGBOURNE

Aston Clo	30 C5
Bere Court Rd	30 A5
Bourne Rd	30 C4
Briars Clo	30 C4
Bucknell Av	30 C4
Chiltern Walk	30 C4
Church St	30 B4
Courtlands Hill	30 A5
Dunluce Gdns	30 C4
Eastfield Rd	30 B2
Flowers Hill	30 B5
Grahame Av	30 C5
Green La	30 A5
Hardwick Rd	30 B2
High St, Pangbourne	30 B4
High St, Whitchurch	30 B2
Horseshoe Rd	30 C4
Kennedy Dri	30 C5
Manor Rd	30 B2
Meadowside Rd	30 B4
Pangbourne Hill	30 A5
Purley Way	30 C4
Reading Rd	30 B4
Riverview Rd	30 B4
St James Clo	30 B4
Shooters Hill	30 A3
Short St	30 C4
Station Rd	30 B4
Sulham La	30 C6
Swanston Field	30 B2
Thames Av	30 B3
The Moors	30 B4
The Square	30 B4
Tidmarsh Rd	30 B6
Whitchurch Bri	30 B3
Whitchurch Rd	30 B4
Wilder Av	30 C4
Woodview Rd	30 C4

PURLEY

Addiscombe Chase	31 A3
Allison Gdns	31 A2
Andover Clo	31 C6
Apple Clo	31 A3
Armour Hill	31 B5
Armour Rd	31 B5
Ashton Clo	31 B6
Avington Clo	31 A6
Back La	31 A5
Barbrook Clo	31 C4
Beatty Dri	31 C6
Beech Rd	31 A2
Beverley Rd	31 B6
Blundells Rd	31 A2
Bowfell Clo	31 A4
Bowling Green La	31 A2
Bracken Clo	31 A5
Brading Way	31 B1
Bradwell Rd	31 D5
Brendon Clo	31 D5
Brierley Pl	31 A3
Brooksby Rd	31 B4

Carlisle Rd	31 C4
Causmans Way	31 B5
Cecil Aldin Dri	31 A2
Chapel Hill	31 A6
Chelsea Clo	31 D6
Chepstow Rd	31 B5
Chestnut Gro	31 C2
Childrey Way	31 A6
Chiltern Vw	31 B2
Church Mews	31 B2
Clanfield Cres	31 B4
Clay Clo	31 A5
Clements Mead	31 A6
Clevedon Rd	31 C4
Coalport Way	31 C6
Colyton Way	31 B2
Condor Clo	31 A3
Coniston Dri	31 D5
Copse Clo	31 B3
Cornwall Clo	31 A3
Cornwell Coppice	31 A5
Cotswold Way	31 A5
Cranmer Clo	31 A4
Crescent Rd	31 B6
Cromer Clo	31 A6
Dark La	31 A5
Dartington Clo	31 D6
Delft Clo	31 D6
Dell Rd	31 B5
Denby Way	31 D6
Derwent Av	31 D5
Devonshire Gdns	31 A3
Downs Way	31 A4
Dresden Way	31 D6
Dudley Clo	31 C5
Dudley Mews	31 C5
Duncan Gdns	31 B2
Dunstall Clo	31 B6
Edenhall Clo	31 C4
Elder Clo	31 A6
Elmstone Dri	31 B6
Elsley Rd	31 C4
Elstree Clo	31 C5
Elyham Rd	31 A2
Fairford Rd	31 A5
Farm Clo	31 B2
Farmilot Clo	31 B2
Felton Way	31 B6
Fern Glen	31 B6
Ferndale Clo	31 C3
Filbert Dri	31 A6
Fircroft Clo	31 B5
Five Acre	31 A6
Forest Hill	31 D5
Fulbrook Cres	31 B3
Gipsy La	31 D5
Glebe Rd	31 A1
Goodliffe Gdns	31 B3
Grasmere Av	31 D4
Greenacre Mt	31 C6
Gwynne Clo	31 B3
Hartslock Way	31 B5
Hawthornes	31 A4
Hazel Rd	31 B2
Hazelwood Clo	31 B6
Highfields Rd	31 A2
Highworth Way	31 A5
Hillview Clo	31 A5
Holford Clo	31 A6
Holkham Clo	31 D6
Home Croft	31 A6
Hornbeam Clo	31 B2
Hornsea Clo	31 C6
Hucklebury Clo	31 B2
Hume Clo	31 A5
Huscarle Way	31 A3
Ivy Bank	31 A6
Juniper Way	31 B4
Kentwood Clo	31 C6
Kentwood Hill	31 C6
Kernshaw Dri	31 A3
Kirkfell Clo	31 A5
Knowsley Rd	31 A4
Lamorna Cres	31 A5
Larissa Clo	31 C5
Larkswood Clo	31 C4
Laytom Rs	31 A4
Lilac Clo	31 B2
Lister Clo	31 A2
Little Heath Rd	31 A6
Little Oaks Dri	31 B6
Long La	31 A3
Longleat Dri	31 A4
Lower Armour Rd	31 C6
Lower Elmstone Dri	31 B6
Lucey Clo	31 A4
Lyme Gro	31 B6
Lytham End	31 A4
Mapledurham Dri	31 B1
Mapledurham Vw	31 DC
Marling Clo	31 B5
Marshall Clo	31 C3
Marten Pl	31 A3
Menpes Rd	31 A3
Mid Winter Clo	31 C6
Minton Clo	31 D6
Myrtle Clo	31 B3
Nevis Rd	31 B4
New Hill	31 B2
Newbery Clo	31 C5
Norcot Rd	31 C6
Nursery Gdns	31 A2
Oak Tree Copse	31 C4
Oak Tree Rd	31 B5
Oak View	31 B6
Oakham Clo	31 C4
Oaktree Walk	31 B2
Old Farm Cres	31 A4
Orchard Clo	31 A3
Oregon Av	31 B3
Overdown Rd	31 A5
Overlanders End	31 D4
Oxford Rd	31 A2
Park Walk	31 B2
Parkhill Dri	31 C4
Pierces Hill	31 B5
Pikeshaw Way	31 B5
Polsted Rd	31 C6
Portmeirion Gdns	31 D6
Pottery Rd	31 D6
Primrose Clo	31 B2
Prince William Dri	31 A5
Purley La	31 A2
Recreation Rd	31 C6
Redwood Way	31 B3
Rissington Clo	31 C4
River Gdns	31 C2
Rodway Rd	31 D5
Romany La	31 D6
Rosemead Av	31 A4
Rydal Av	31 D5
Sage Rd	31 B3
St Marys Av	31 B2
Sandgate Av	31 D5
Scafell Clo	31 A5
School Rd	31 C6
Sheraton Dri	31 A6
Sherwood Rise	31 A2
Shipton Clo	31 B5
Simons Clo	31 B3
Skerritt Way	31 B3
Skilton Rd	31 B4
Southerndene Clo	31 B5
Spode Clo	31 D6
Staddle Stone Clo	31 A4
Staffordshire Clo	31 D6
Swansea Ter	31 C5
Swinbrook Clo	31 B4
Swiss Cottage Clo	31 A6
Talbot Way	31 A3
Thames Reach	31 B2
The Beeches	31 C4
The Cedars	31 B5
The Glade	31 B3
The Holt	31 B3
The Hydes	31 A3
The Knoll	31 A5
The Short	31 B2
Theobald Rd	31 B3
Thirlmere Av	31 D5
Thistledown	31 A5
Tilling Clo	31 A5
Trelawney Dri	31 A5
Trenthams Clo	31 A2
Tring Rd	31 B4
Tuscan Clo	31 D6
Ullswater Dri	31 C4
Vale Cres	31 D5
Ventnor Gro	31 B6
Victoria Rd	31 B6
Wandhope Way	31 A5
Warbreck Dri	31 A3
Wardle Av	31 B5
Warley Rs	31 A3
Waterside Dri	31 B2
Weald Rise	31 D5
Wealden Way	31 D6
Wedgewood Way	31 D6
Welland Clo	31 A6
Western Oaks	31 C4
Westridge Av	31 A2
Westwood Glen	31 A6
Westwood Rd	31 B5
Westwood Row	31 B5
White Lodge Clo	31 A3
Winston Way	31 A2
Wintringham Way	31 B1
Wyre Ct	31 A3

READING

Abattoirs Rd	32 D2
Abbey Sq	33 E3
Abbey St	33 E3
Abbots Walk	33 E3
Acacia Rd	33 F4
Addington Rd	33 G5
Addison Rd	32 D1
Admirals Ct	32 D5
Albany Rd	32 B3
Albion Pl	33 F4
Albury Clo	32 A1
Aldworth Clo	32 A5
Alexandra Rd	33 G4
Alfred St	32 D3
Allcroft Rd	33 F5
Alpine St	33 E5
Ambrose Pl	32 C3
Amity Rd	33 H3
Amity St	33 G3
Anglers Way	33 G3
Anstey Rd	32 D3
Argyle Rd	32 B3
Argyle St	32 C3
Armadale Ct	32 A4
Arthur Pl	33 G3
Ashley Rd	32 B5
Audley St	32 B2
Avebury Sq	33 G5
Aveley Walk	33 E5
Avon Pl	33 G3
Baker St	32 C2
Barn Clo	32 A6
Barnstaple St	33 E4
Barnwood Clo	32 B2
Barnwood Ct	32 B2
Baron Ct	32 B3
Barry Pl	32 D2
Basingstoke Rd	33 E6
Bath Rd	32 A5
Battle St	32 C3
Baydon Dri	32 C5
Bedford Rd	32 C3
Belle Vue Ter	32 B4
Bellevue Rd	32 C3
Belmont Rd	32 B3
Bembridge Pl	33 F3
Benyon Ct	32 B4
Benyon Ct Mews	32 B4
Beresford Rd	32 B3
Berkeley Av	32 B4
Berkeley Ct	32 C4
Betam Rd	33 G3
Bexley St	32 A4
Blagrave St	33 E2
Blakes Cotts	33 F3
Blenheim Gdns	33 H4
Blenheim Rd	33 G3
Blythe Walk	33 E5
Body Rd	32 D3
Boston Av	32 C5
Boult St	33 F4
Boults Walk	33 E5
Bourne Av	33 E6
Bridge St	33 E3
Brigham Rd	32 D1
Britten Rd	33 E6
Broad St	32 D3
Brook St West	32 D4
Broughton Clo	32 A1
Brownlow Rd	32 C4
Brunswick Hill	32 B3
Brunswick St	32 B4
Bulmershe Rd	33 H4
Burfield Ct	32 D2
Butter Market	33 E3
Cambridge St	32 C2
Canal Way	33 G3
Canford Ct	32 A3
Cannon St	32 C3
Cardiff Rd	32 C1
Cardiff Rd Ind Est	32 C1
Cardigan Gdns	33 H5
Cardigan Rd	33 H4
Cardinal Clo	33 E1
Carey St	32 D3
Carnarvon Rd	33 H4
Caroline Ct	32 C4
Caroline St	32 C2
Carsdale Clo	32 C5
Castle Cres	32 C4
Castle Hill	32 C4
Castle St	32 D3
Catherine St	32 B2
Caversham Bri	32 D1
Caversham Rd	32 D1
Chain St	33 E3
Champion Rd	33 F1
Chancery Mews	32 C3
Charles St	32 C3
Charndon Clo	33 E5
Chatham St	32 C3
Cheapside	32 D3
Chester St	32 A2
Chesterman St	33 E4
Cholmeley Pl	33 H3
Cholmeley Rd	33 G2
Cholmeley Ter	33 H3
Christchurch Gdns	33 F5
Christchurch Rd	33 F5
Church St	33 E4
Cintra Av	33 F6
Circuit La	32 A6
Clayton Walk	33 F6
Clent Rd	33 E6
Clifton St	32 C3
Coldicut St	33 F1
Coley Av	32 C4
Coley Hill	32 D4
Coley Park Rd	32 C4
Coley Pl	32 D4
Collis St	33 E5
Connaught Clo	32 B3
Connaught Rd	32 B4
Coventry Rd	33 H3
Cow La	32 C1
Cowper Way	32 A6
Cranbury Rd	32 A4
Craven Rd	33 G4
Cremyll Rd	32 C1
Crescent Rd	33 H4
Cross St	33 E3
Crossland Rd	33 E4
Crown Pl	33 F4
Crown St	33 E4
Cumberland Rd	33 G2
Curzon St	32 B2
Dale Rd	33 E5
De Beauvoir Rd	33 G3
De Bohun Rd	33 E2
De Montfort Rd	33 E1
Deans Rd	33 G1
Deansgate Rd	33 E4
Deepdene Clo	32 C3
Denmark Rd	33 G4
Derby St	32 C2
Donnington Gdns	33 G4
Donnington Rd	33 G3
Dorchester Ct	32 A4
Dorothy St	33 E4
Dorset St	32 A2
Dover St	32 D4
Downshire Sq	32 C3
Duke St	33 E3
Dusseldorf Way	32 D3
Earley Pl	33 E3
East St	33 E3
Eastern Av	33 H3
Eaton Pl	32 D3
Edenham Cres	32 B5
Edgehill St	33 E5
Edinburgh Rd	32 B3

Name	Ref
Eldon Pl	33 F3
Eldon Rd	33 F3
Eldon Sq	33 F3
Eldon St	33 G3
Eldon Ter	33 F3
Elgar Rd	33 E4
Elgar Rd Sth	33 E5
Elizabeth Walk	33 E5
Elm La	33 C5
Elm Lodge Av	32 A3
Elm Park Rd	32 A3
Elmhurst Rd	33 G5
Epping Clo	32 C4
Epsom Ct	32 C4
Erleigh Rd	33 G4
Essex St	33 E5
Faircross Rd	32 A6
Fatherson Rd	33 G3
Fawley Rd	32 A5
Field Rd	32 D4
Filey Rd	33 H3
Florence Ct	32 A4
Florence Walk	33 G3
Florida Ct	32 C4
Fobney St	32 D4
Forbury Ind Est	33 F2
Forbury Rd	33 E2
Forge Clo	33 F1
Foundry St	33 E4
Foxhill Rd	33 H4
Francis St	33 E4
Franklin St	32 C3
Freshwater Rd	33 H3
Friar St	32 D3
Froxfield Av	32 C5
Fulham Rd	33 E4
Fulmead Rd	32 A2
Garnet Hill	32 D4
Garnet St	32 D4
Garrard St	32 D2
Gas La	33 E3
Gas Works Rd	33 F3
George St	32 C3
George St	33 E1
Glebe Rd	33 F5
Glenbeigh Ter	32 B4
Gloucester Rd	32 A3
Goldsmid Rd	32 C3
Gordon Pl	32 A2
Gosbrook Rd	33 F1
Gower St	32 C3
Granby St	33 H3
Granby Gdns	33 H3
Great Knollys St	32 C2
Greyfriars Rd	32 D2
Gun St	33 E3
Hagley Rd	33 E6
Hall Pl	33 E3
Hamilton Rd	33 H3
Harrow Ct	32 C4
Hart St	32 C3
Hatherley Rd	33 G4
Hay Rd	32 C5
Haywood Ct	33 H3
Henry St	33 E4
Heron Island	33 F1
Heron Way	32 B6
High St	33 E3
Highgrove St	33 F5
Highgrove Ter	33 F5
Hilcot Rd	32 A3
Hill St	33 E4
Hodsoll Rd	32 C2
Hollins Walk	32 A4
Holybrook Rd	32 D5
Hosier St	32 D3
Howard St	32 D3
Hungerford Dri	32 B5
Huntley Ct	33 H4
Inglewood Ct	32 A5
James St	32 C3
Janson Ct	32 C4
Jesse Ter	32 D3
Johannes Ct	32 B4
Josephine Ct	32 B4
Jubilee Sq	33 E4
Junction Rd	33 H4
Katesgrove La	33 E4
Kearsley Rd	32 A4
Kenavon Dri	33 F2
Kendrick Rd	33 F4
Kenilworth Av	32 A6
Kennet Side	33 F3
Kennet St	33 F3
Kensington Rd	32 A3
Kent Rd	32 A3
Kimberley Clo	32 B5
King St	33 E3
Kingfisher Pl	33 E2
Kings Meadow Rd	33 E2
Kings Rd	33 E1
Kings Rd	33 E3
Kings Wood Clo	32 B4
Kingsgate Pl	33 G3
Kingsgate St	33 G3
Kinver Walk	33 E5
Lancaster Clo	33 F5
Lancing Rd	32 B3
Leopold Walk	33 G3
Lesford Rd	32 B6
Lesley Ct	32 B4
Letcombe St	33 E4
Liebenrood Rd	32 A4
Lima Ct	32 C4
Lincoln Rd	33 F6
Little Johns La	32 A2
Little St	32 C2
Littlecote Dri	32 C5
Liverpool St	33 H2
Livery Clo	33 F3
Lock Pl	33 G3
London Rd	33 E4
London St	33 E3
Lorne St	32 C3
Loverock Rd	32 A1
Lower Brook St	32 D4
Lower Field Rd	32 D4
Lundy La	32 A3
Lydford Rd	33 G5
Lynden Mews	33 E6
Lynmouth Rd	33 E1
Maitland Rd	32 C4
Maldon Clo	32 B4
Mallard Row	32 D4
Malthouse La	32 C3
Maltings Pl	32 D3
Malvern Ct	33 G5
Manchester Rd	33 H3
Mansfield Rd	32 C4
Maria Ct	32 B4
Market Pl	33 E3
Market Way	33 E3
Marlborough Av	33 G5
Marlborough Ct	32 C4
Marsh Ct	32 A3
Mason St	32 C3
Meadow Rd	32 D1
Mellor Walk	33 E5
Merchants Pl	33 E2
Milford Rd	32 C2
Mill Grn	33 F1
Mill La	33 E3
Mill Rd	33 F1
Millman Rd	33 E5
Minster St	33 E3
Mitcham Clo	33 E5
Monkley Ct	33 F1
Monks Way	32 A5
Montague St	33 G3
Morgan Rd	33 F5
Mount Pleasant	33 E4
Mount Pleasant Gro	33 E4
Mount St	33 E5
Muirfield Clo	33 G3
Mundesley St	33 E4
Napier Rd	32 E2
New Bright St	32 D4
New Rd	33 G5
Newark St	33 E4
Newcastle Rd	33 F6
Newport Rd	32 D1
Nimrod Way	33 E6
Norfolk Rd	32 A3
North St	32 D2
Northcourt Av	33 G6
Northfield Cotts	32 D2
Northfield Rd	32 D2
Northumberland Av	33 F6
Norton Rd	33 H3
Norwood Rd	33 G3
Oaklands	33 H4
Orchard St	33 E4
Ormsby St	32 C3
Orts Rd	33 F3
Osterley Ct	33 H4
Oxford Rd	32 A2
Paddock Rd	33 G1
Paddock Rd Ind Est	33 G1
Park House La	32 A4
Parkside Rd	32 A4
Parthia Clo	33 E4
Patriot Pl	33 F3
Pelham Ct	32 A4
Pell St	33 E4
Penroath Av	32 A4
Pepper La	33 H6
Petworth Ct	32 B4
Piggotts Rd	33 F1
Portman Rd	32 A2
Portway Ct	32 C4
Preston Rd	33 E5
Prince of Wales Av	32 A3
Princes St	33 F3
Prospect St	32 C3
Queen Victoria St	33 E3
Queens Cotts	33 F3
Queens Lawn	33 G5
Queens Rd	33 E1
Queens Rd	33 E3
Queens Walk	32 D3
Radstock Rd	33 H3
Randolph Rd	32 D1
Reading Bri	33 E2
Redlands Rd	33 F4
Redlane St	33 G5
Regent St	33 H3
Restwold Clo	32 A6
Richfield Av	32 C1
Richmond Rd	32 A2
River Rd	33 E4
Riverside Ct	33 H2
Rose Kiln La	32 D5
Rose Walk	32 D4
Ross Rd	32 D1
Rowley Rd	33 E6
Rupert St	33 G3
Russell St	32 C4
Rutland Rd	32 A3
Rylstone Rd	32 A2
Sackville St	32 D2
St Bartholomews Rd	33 H3
St Giles Clo	33 E4
St Johns Hill	33 F3
St Johns Rd	33 F3
St Johns St	33 F3
St Marys Butts	32 D3
St Saviours Rd	32 D5
St Stephens Clo	33 E1
Salisbury Rd	32 B3
Sarum Ct	32 B4
School Ter	33 G2
Send Rd	33 F1
Shaw Rd	32 C5
Shenstone Rd	33 E6
Shepley Dri	32 A6
Sherman Rd	33 E4
Sherwood St	32 A2
Shinfield Rd	33 G6
Shireshead Clo	32 A4
Short St	33 E4
Sidmouth St	33 F3
Silchester Rd	32 A6
Silver St	33 E4
Simmonds St	32 D3
Somerston Ct	32 C3
South St	33 E3
Southampton St	33 E4
Southcote Farm La	32 A5
Southcote La	32 A5
Southcote Rd	32 B4
Southern Hill	33 G5
Spring Gro	33 F4
Spring Ter	33 E5
Stanley Gro	32 C3
Stanley St	32 C2
Stanshawe Rd	32 D2
Stapleford Rd	32 A6
Star La	33 E3
Star Rd	33 F1
Station Hill	32 D2
Station Rd	33 E2
Suffolk Rd	32 A3
Sun St	33 G3
Surrey Rd	33 E6
Sutherlands Av	33 F5
Sutton Walk	33 F5
Suttons Bus. Park	33 H2
Swainstone Rd	33 F6
Swan Pl	32 D4
Swansea Rd	32 D2
Sylvan Walk	32 A6
Talbot Ct	32 D4
Tallis La	32 A6
Taylor Ct	32 C3
Tazewell Ct	32 C4
Temple Pl	32 D4
Tessa Rd	32 C1
Tetbury Ct	32 C3
Thames Av	32 D1
Thames Side	32 D1
The Chancellors Way	33 G6
The Dell	33 F4
The Firs	32 C4
The Forbury	33 E2
The Grove	33 F3
The Mews	33 H4
The Mount	33 F5
The Oracle	33 E3
The Parade	32 A6
The Queens Dri	33 G6
Thorn La	33 E3
Thorn St	32 D3
Thornton Mews	32 A1
Thornton Rd	32 A2
Tilehurst Rd	32 A4
Tintern Cres	32 C5
Tippett Rise	33 E5
Tofrek Ter	32 A3
Town Pl	33 F3
Trafalgar Ct	32 B4
Trafford Rd	32 C1
Trelleck Rd	32 C5
Treyarnon Ct	33 H4
Trinity Pl	32 C3
Tudor Rd	32 D2
Tupsley Rd	32 C5
Tyberton Pl	32 C5
Union St	33 E3
Upavon Dri	32 C5
Upper Crown St	33 E4
Upper Redlands Rd	33 G5
Vachel Rd	32 D2
Valentia Rd	32 B2
Valerie Ct	32 C4
Valpy St	33 E2
Vastern Rd	33 E1
Vicarage Rd	33 F5
Victoria Mews	32 A4
Victoria Sq	33 G3
Victoria St	33 G3
Victoria Way	33 G3
Waldeck St	33 E5
Walkers Pl	32 A3
Wantage Rd	32 A3
Warwick Rd	33 F6
Waterloo St	33 E5
Waterside Gdns	32 D3
Watlington St	33 F3
Waverley Ct	32 B4
Waybrook Cres	33 H5
Waylen St	32 C3
Weldale Pl	32 D2
Weldale St	32 D2
Wellington Av	33 G6
Wensley Rd	32 B6
West Fryerne Ct	32 A4
West Hill	33 E4
West St	32 D3
Westbrook Rd	32 A2
Westcote Rd	32 A4
Western Elms Av	32 B4
Western Rd	32 B4
Whiley Park La	33 G6
Whitby Dri	33 F5
Whiteknights Rd	33 H5
Whitley St	33 E5
William St	32 C2
Willow St	32 D4
Wilson Rd	32 A3
Wilton Rd	32 A3
Winchester Rd	33 E6
Wingrove Rd	32 A4
Winser Dri	32 A6
Wiston Ter	32 E2
Wokingham Rd	33 H3
Wolseley St	32 D4

Street	Ref
Wolsey Rd	33 E1
Wood Green Clo	32 B3
Woodstock St	33 G3
Yield La	33 E3
York Rd	32 D2
Zinzan St	32 D3

SANDHURST

Street	Ref
Abingdon Rd	35 E3
Acacia Av	35 E3
Ackrells Mead	34 C1
Albion Rd	34 D4
Allenby Rd	35 H6
Allendale Clo	34 B2
Alton Rd	34 D6
Ambarrow Cres	34 B3
Ambarrow La	34 A2
Andover Pl	34 D6
Apple Tree Way	35 E3
Atrebatti Rd	35 E3
Avocet Cres	35 F4
Bacon Clo	35 E6
Balintore Ct	35 E5
Balliol Way	35 F3
Beaulieu Gdns	34 D6
Beech Ride	34 D3
Beechnut Dri	34 C6
Bernersh Clo	35 E3
Berrybank	35 F6
Birdwood Rd	35 H5
Birkbeck Clo	35 F4
Bittern Clo	35 F4
Blackbird Clo	35 F4
Blackcap Pl	35 F4
Bluethroat Clo	35 F4
Branksome Hill Rd	35 F5
Braye Clo	35 E3
Breach La	34 C4
Brittain Ct	34 E5
Brook Clo	35 F3
Brookside	34 D5
Broomacres	34 C4
Bullfinch Clo	35 F4
Burghead Clo	35 F5
Burley Way	35 E6
Burne-Jones Dri	35 E6
Camborne Clo	34 B3
Cambridge Rd	35 F3
Carrick La	34 A6
Castlecraig Ct	35 E5
Caves Farm Clo	34 B4
Cedar Clo	34 B4
Centurion Clo	35 E4
Chaffinch Clo	35 F4
Chelwood Dri	34 B3
Cherry Tree Clo	35 E3
Cheviot Rd	34 B3
Chiltern Rd	34 B2
Christchurch La	34 D6
Church Rd, Little Sandhurst	34 B3
Church Rd, Owlsmoor	35 F3
Clarke Cres	35 G5
Cock-A-Dobby	34 C3
College Cres	35 G4
College Rd	35 F5
Compton Clo	34 D3
Constable Way	35 F6
Cookham Clo	34 D3
Copperfield Rd	35 F2
Cormorant Clo	35 F4
Cornbunting Clo	35 F4
Cotswold Rd	34 B3
Crake Pl	35 F4
Crane Ct	35 F4
Cricket Hill Rd	34 A6
Crofters Clo	34 C4
Crown Pl	35 F3
Crowthorne Rd	34 C4
Cruikshank Lea	35 F6
Culver Clo	35 E3
Dale Gdns	34 C4
Darby Green La	34 C6
Darleydale Clo	35 F2
Devon Clo	35 E5
Dovedale Clo	35 F2
Durham Clo	35 F2
Eagles Nest	34 B3
Edgbarrow Rise	34 C2
Egerton Rd	35 G6
Evenlode Way	35 E4
Evesham Walk	35 E4
Fairmead Clo	35 F5
Fakenham Way	35 E3
Faringdon Clo	34 D3
Farncrosse Clo	34 D4
Faversham Rd	35 E3
Ferryhill Dri	34 B3
Fielding Clo	35 F6
Findhorn Clo	35 F5
Firtree Clo	34 B3
Florence Rd	35 F5
Forbes Clo	35 E5
Forest End Rd	34 B3
Fortress Clo	35 F5
Fraser Mead	35 F6
Frodsham Way	35 F2
Frys La	34 A5
Georgeham Rd	35 E2
Gibbons Clo	34 D4
Girton Clo	35 G3
Glen Innes	35 G3
Gordon Walk	34 A6
Goughs Meadow	34 C5
Governors Rd	35 G6
Grampian Clo	34 B2
Grantham Clo	35 F3
Green La	34 D4
Greenways	34 C3
Grove Cres	35 G6
Haig Rd	35 G6
Hailsham Clo	35 E3
Hancombe Rd	34 B2
Hartley Clo	34 D6
Hartsleap Clo	34 C3
Hartsleap Rd	34 C4
Harvard Rd	35 G3
Hatherwood	34 A6
Hearsey Gdns	34 A5
Hexham Clo	35 E2
High St, Little Sandhurst	34 B3
High St, Sandhurst	34 A3
Hillside	35 G5
Hogarth Clo	35 F6
Hone Hill	34 C4
Hopeman Clo	35 E4
Hornbeam Clo	35 E3
Horsham Rd	35 E3
Humber Way	35 E4
Hungerford Clo	34 D4
Inverness Way	35 E5
Isis Way	35 E4
Jennys Walk	34 A6
Keble Way	35 G2
Kevins Dri	34 A5
Keynsham Way	35 E2
Kilmuir Clo	35 E5
Kings Keep	34 D3
Kings Walk	35 G6
Kirkham Clo	35 E2
Landseer Clo	35 F6
Larkswood Clo	34 C3
Laundry La	35 F6
Lewisham Way	35 E2
Lindale Rd	35 F2
Little Moor	34 D3
Lodge Gro	34 A6
Long Mickle	34 C3
Longdown Rd	34 C3
Lower Church Rd	34 A3
Lower Wokingham Rd	34 A1
Lowry Clo	35 F6
Lych Gate Clo	34 A4
Lyndhurst Av	34 D6
Magdalena Rd	35 G2
Magnolia Clo	35 E3
Maple Clo	34 B3
Marshall Rd	35 E5
Mason Pl	34 B4
Maxine Clo	34 C3
May Clo	35 E3
Maybrick Clo	34 A3
Melksham Clo	35 E2
Merton Clo	35 G2
Mickle Hill	34 C3
Millins Clo	35 F3
Moffatts Clo	34 C4
Montgomery Clo	34 D4
Moor Clo	35 F3
Moray Av	35 E4
Mount Pleasant	34 B2
Mountbatten Clo	34 B3
Mulberry Clo	35 E3
Munnings Dri	35 F6
New Rd	34 B4
New Town Rd	34 C4
Nightingale Gdns	34 C4
Nuffield Dri	35 G3
Oak Av	35 E3
Oaktree Way	34 C3
Ockingham Clo	35 E3
Orchard Gate	34 D4
Owlsmoor Rd	35 E4
Oxford Rd	35 F2
Park Dri	34 A6
Park Rd	34 D5
Parsons Field	34 D4
Peddlars Gro	34 A6
Peterhouse Clo	35 G2
Pine Clo	35 G5
Pinehill Rise	34 D4
Pinehill Road	35 E1
Pond Croft	34 A6
Potley Hill	34 A6
Primrose Way	34 D3
Prince Dri	34 B3
Rackstraw Rd	35 E3
Raeburn Way	35 E6
Range Vw	35 G4
Rawlinson Rd	35 H6
Reading Rd	34 A6
Rectory Clo	34 B4
Regents Pl	34 D4
Reynolds Grn	35 E6
Richmond Rd	35 F4
Ringwood Rd	34 D6
Ripplesmore Clo	34 C4
Roberts Rd	35 H6
Robin La	34 D4
Rockfield Way	35 E4
Romsey Clo	35 E6
Rookwood Av	35 F2
Rosedene La	35 F6
Rosemary La	34 D6
Rother Clo	35 E4
Rugby Clo	35 F3
Ryan Mt	34 B4
St Helens Cres	34 C4
St Helens Dri	34 C4
St Johns Rd	34 D5
St Marys Clo	34 D4
St Michaels Rd	34 B4
Sandhurst La	34 C6
Sandhurst Rd	34 C3
Sandhurst Rd, Yateley	34 A6
Sandy La	34 B3
School Hill	34 B3
Scotland Hill	34 C4
Selbourne Clo	35 E6
Severn Clo	35 E4
Shaw La	34 C5
Shrivenham Clo	35 E4
Silver Hill	35 F4
Somerville Cres	34 A6
Sonning Clo	35 E4
South Meadow	35 F1
South Rd	35 F1
Southampton Clo	35 E6
Spring Woods	34 D3
Squirrel Clo	34 C4
Steerforth Copse	35 F2
Sun Ray Est	34 B4
Sycamore Clo	34 C4
Sylvan Ridge	34 C3
Tarbat Ct	35 E4
Templar Clo	34 B4
The Broadway	34 D5
The Close	35 F4
The Nook	34 B4
The Square	35 G6
Theal Clo	35 E4
Thibet Rd	34 D4
Tottenham Walk	35 E3
Travis La	34 D5
Trinity	35 G2
Trotwood Clo	35 F2
Turner Clo	35 F6
Union Clo	35 F2
Uplands Clo	34 C4
Vacob Rd	35 H5
Vale Clo	35 G2
Victoria Rd	35 F3
Vulcan Clo	34 C5
Vulcan Way	34 C5
Wadham	35 G3
Wantage Rd	35 E4
Wargrove Dri	35 E4
Warren Clo	34 C4
Wasdale Clo	35 E2
Wellington Clo	35 E4
Wellington Rd	34 D4
Wellington Ter	35 E4
Westbourne Rd	35 F5
Weybridge Mead	34 A5
Whistler Gro	35 E6
White Rd	35 G6
Whitmore Clo	35 E3
Whittle Clo	34 B3
Willow End	34 C4
Willow Way	34 B3
Winchester Way	34 D6
Windrush Heights	34 C4
Woodbine Clo	34 D5
Woodside	35 G5
Yateley Rd	34 A4
Yeovil Rd	35 F3
York Town Rd	34 C5
York Way	34 C4

SHINFIELD

Street	Ref
Aborfield Rd	42 C2
Babbington Rd	42 B1
Brookers Hill	42 B1
Chestnut Cres	42 C2
Church La	42 A1
Cutbush La	42 C1
Drewett Clo	42 A1
Fairmead Rd	42 C2
Farmers Clo	42 A1
Goddard Clo	42 C1
Hirtes Av	42 C2
Hollow Hill	42 C1
Hyde End La	42 A2
Hyde End Rd	42 C2
Ilbury Clo	42 C1
Kendal Av	42 C1
Lane End Clo	42 C1
Lexington Grn	42 A1
Leyland Gdns	42 C1
Longmoor Rd	42 B1
Lower Earley Way	42 A1
Millworth La	42 C2
Milson Clo	42 C1
Oatlands Rd	42 C2
Old Whitley Wood La	42 A1
Pattinson Rd	42 B1
Pendred Rd	42 A1
Rosecroft Way	42 C1
Salmond Rd	42 A1
Sampage Clo	42 A1
Sandleford Clo	42 A1
Seymour Av	42 C1
Shinfield Rd	42 C1
Swallowfield Dri	42 A1
Swallowfield Grn	42 A1
Vickers Clo	42 C2
Wheatfields Rd	42 C1
Whitley Wood La	42 A1
Wychelm Rd	42 C2

SLOUGH

Street	Ref
Adelphi Gdns	36 B4
Alan Way	37 H2

Name	Ref		Name	Ref		Name	Ref		Name	Ref
Albert Clo	36 D5		Dashwood Clo	37 F6		Ladyday Pl	36 A4		St Bernards Rd	37 F6
Albert St	36 C5		Datchet Rd	36 D6		Lake Av	36 A3		St Johns Rd	36 D3
Albion Clo	36 D4		Dawes Moor Clo	37 G2		Lambert Av	37 H5		St Laurence Way	36 D5
Aldborough Spur	36 B2		Devon Av	36 A1		Langley Rd	37 F5		St Marys Rd	37 H4
Aldin Av N	37 E4		Diamond Rd	37 E4		Lansdown Av	36 B4		St Pauls Av	36 D3
Aldin Av S	37 E5		Diana Clo	37 H2		Lansdown Ct	36 C4		Salt Hill Av	36 A4
Alexandra Rd	36 B5		Dolphin Rd	37 E4		Lascelles Rd	37 E6		Salt Hill Dri	36 A3
Almonds Way	37 E1		Dornels	37 F2		Laurel Av	37 H5		Salt Hill Way	36 A3
Alpha St	36 D5		Downs Rd	37 G5		Ledgers Rd	36 B5		Second Cres	36 A1
Amanda Ct	37 G5		Dudley Ct	37 D6		Leeds Rd	36 B3		Sermed Ct	37 F3
Appletree La	37 F5					Leiston Spur	36 B1		Seymour Rd	36 A4
Arborfield Clo	36 C5		Eastfield Clo	36 D5		Lerwick Spur	36 B1		Shackleton Rd	36 C3
Arthur Rd	36 B4		Edmunds Way	37 F1		Litcham Spur	36 B2		Shaggy Calf La	37 D2
Arundel Ct	37 G6		Elliman Av	36 C2		Littledown Rd	36 C3		Sheey Way	37 F2
August End	37 H1		Ellis Av	36 B4		Littleport Spur	36 B2		Sheffield Rd	36 A2
Australia Rd	37 E4		Elmwood Rd	37 E3		Locke Gdns	37 F5		Sherman Rd	36 C1
Aylesbury Cres	36 B2		Ely Av	36 A1		Lodden Spur	36 B2		Sherwood Clo	37 H6
Azalea Way	37 H2		Emerald Ct	36 C5		London Rd	37 E5		Shortfern	37 F2
			Everard Av	36 B4		Lynwood Av	37 G6		Slough Rd	36 B6
Bannister Clo	37 H5								Snape Spur	36 B1
Bath Rd	36 A4		Farm Cres	37 F1		Mackenzie St	36 C4		South Green	36 C3
Baylis Rd	36 B3		Farm La	36 B2		Magnolia Gdns	37 F6		Spackmans Way	36 A6
Beech Rd	37 H5		Farnham Rd	36 A1		Malpas Rd	37 E3		Springate Field	37 H5
Beechwood Gdns	36 C5		First Cres	36 A1		Malton Av	36 A1		Stanley Cotts	36 C4
Belfast Av	36 A2		Fleetwood Rd	36 D3		Mansell Clo	37 E1		Starwood Ct	37 F5
Belgrave Pl	37 E5		Fox Rd	37 G6		Maple Cres	37 F3		Statfield Rd	36 D4
Belgrave Rd	36 C3		Foxherne	37 F5		Martin Rd	36 B5		Stewart Av	36 C1
Bell Clo	37 F1					Maryside	37 H5		Stile Rd	37 G6
Bellvue Pl	36 D6		Gatewick Clo	36 C4		Meadow Rd	37 H6		Stoke Gdns	36 C3
Benson Clo	36 D3		George Green Rd	37 H2		Melbourne Av	36 A2		Stoke Poges La	36 B4
Beresford Av	37 F3		Gilliat Rd	36 C3		Mere Rd	36 D6		Stoke Rd	36 C4
Berryfield	37 F2		Gilmore Clo	37 F4		Merton Rd	36 D5		Stokesay	36 D2
Beverley Ct	37 E5		Glanmor Rd	37 F3		Middle Green	37 H3		Sumburgh Spur	36 B1
Birley Rd	36 B2		Glenavon Gdns	37 G6		Middlegreen Rd	37 G5		Sussex Clo	37 E5
Bishops Rd	36 D5		Glentworth Pl	36 A4		Mildenhall Rd	36 B2		Sussex Pl	37 E5
Blackthorn Dell	37 F5		Gloucester Av	36 A1		Mill St	36 D3		Sutton Av	37 F5
Blair Rd	36 B4		Godolphin Rd	36 B2		Mina Av	37 G5		Sycamore Walk	37 H2
Blandford Clo	37 G5		Goodmans Pk	37 F4		Mirador Cres	37 E2		Sydney Gro	36 A2
Blandford Ct	37 G6		Gosling Grn	37 H6		Moat Dri	37 F1			
Blandford Rd N	37 G5		Gosling Rd	37 H5		Montague Rd	36 C3		The Cherries	37 F2
Blandford Rd S	37 G6		Grafton Clo	37 H2		Montem La	36 A4		The Crescent	36 B5
Blinco La	37 H1		Grangewood	37 F1		Moreau Walk	37 H2		The Drive	37 H5
Borderside	37 E2		Grant Av	36 C2		Mortimer Rd	37 H6		The Frithe	37 E2
Boston Gro	36 A1		Granville Av	36 A1		Mountbatten Clo	36 D5		The Glen	37 F6
Bourne Rd	36 A5		Grasmere Av	36 D3		Muddy La	36 C1		The Green	36 A5
Brackenforde	37 F5		Grays Pl	36 D3		Mundesley Spur	36 B2		The Grove	36 D5
Bradley Rd	36 B3		Grays Rd	36 C3		Myrtle Cres	36 D2		The Link	37 F2
Brammas Clo	36 A5		Greendale Mews	36 D3					The Mews	36 C6
Bristol Way	36 C4		Grove Clo	36 D5		New Sq	36 C4		The Normans	37 E1
Broad Platts	37 G6					Newberry Way	36 B5		Third Cres	36 A1
Broadmark Rd	37 E3		Hadlow Ct	36 A3		Newnham Clo	36 D3		Thurston Rd	36 B1
Brunel Way	36 C4		Halking Croft	37 F5		Nixey Clo	37 E5		Trelawney Av	37 G6
Bryant Av	36 C1		Hampshire Av	36 A1		Norfolk Av	36 A1		Tunis La	36 A4
Buckingham Av E	36 A2		Hanover Clo	36 D6		North Green	36 C3		Turner Rd	37 F5
Buckingham Gdns	36 C5		Harewood Pl	37 E5		Northampton Av	36 A1		Turton Way	36 B5
Buckland Av	37 E6		Hartland Clo	36 B4		Norway Dri	37 E1			
Burlington Av	36 C5		Hatfield Rd	36 D5		Nursery La	37 G4		Upton Clo	36 C6
			Hawthorn Cres	36 C2					Upton Court Rd	36 D6
Canada Rd	37 E4		Hawtrey Clo	37 F5		Oakley Cres	36 B2		Upton Pk	36 C5
Carlisle Rd	36 A3		Hazlemere Rd	37 E3		Oatlands Dri	36 A1		Upton Rd	36 D6
Carlton Rd	37 E3		Hempson Av	37 F5		Osborne St	36 C5		Uxbridge Rd	37 E4
Carmarthen Rd	36 B3		Hencroft St	36 D5						
Carrington Rd	36 C3		Henry Rd	36 B4		Palmerston Av	37 F5		Vale Gro	36 C6
Castle St	36 D6		Hermitage Clo	37 F5		Park La	37 E6		Vicarage Pl	37 E5
Chalvey Gdns	36 C5		Herschel St	36 C5		Park St	36 D5		Victoria Rd	37 F4
Chalvey Pk	36 C5		High St	37 D5		Paxton Av	36 A5		Victoria St	36 D5
Chalvey Rd E	36 B5		High St, Chalvey	36 A5		Petersfield Av	37 D3			
Chalvey Rd W	36 B5		Hillary Rd	37 H5		Pitts Rd	36 A3		Wallis Clo	37 E5
Chapel St	36 D5		Hillersdon	37 E1		Plough Lees La	36 C3		Warren Clo	37 H6
Cherry Av	37 G5		Hillside	36 B5		Post Office La	37 G2		Warrington Av	36 A1
Chester Rd	36 B1		Hilperton Rd	36 C4		Preston Rd	37 F3		Waterbeach Rd	36 A1
Chestnut Av	37 H5		Holmedale	37 F2		Prestwood	37 F2		Waverley Rd	36 A1
Chichester Ct	37 E5		Homewood	37 G2		Princes St	37 E4		Webb Clo	37 H6
Church La	37 F1		Hornbeam Gdns	36 D6		Priors Clo	36 D5		Welden	37 F2
Church St	36 C5		Hubert Rd	37 H6					Wellesley Rd	36 D4
Church St, Chalvey	36 A5		Hughendon Rd	36 B2		Quaves Rd	37 F5		Wellington St	36 C4
Churchfield Mews	37 E1		Hurworth Rd	37 F6		Queens Ct	36 D3		Westfield La	37 H2
Cippenham La	36 A4					Queens Rd	36 C3		Wexham Rd	37 E1
Clements Clo	37 E5		India Rd	37 E4		Queensmere Rd	36 D4		Wheatland Rd	37 F6
Clifton Rd	37 E5					Quinbrooks	37 F2		Whitby Rd	36 A2
Clive Ct	36 A5		John Taylor Ct	36 A4					White Clo	36 B3
Cockett Rd	37 H5					Ragstone Rd	36 B5		White Hart Rd	36 A6
Coftards	37 G2		Kaywood Clo	37 G6		Railway Ter	36 C4		Whiteford Rd	36 C1
College Av	36 B5		Keel Dri	36 A5		Rambler La	37 G6		Whitehaven	36 D3
Colonial Rd	37 E4		Kendal Clo	36 D3		Randolph Rd	37 H6		Whittenham Clo	37 D3
Concorde Way	36 A5		Kendal Dri	36 D3		Ravensfield	37 G4		William St	36 C4
Conegar Ct	36 B4		Kendrick Rd	37 F6		Red Ct	36 B4		Windmill Rd	36 A4
Connaught Rd	37 E4		Kenilworth Clo	36 D6		Redcott Ms	37 F6		Windsor Rd	36 C6
Coronation Av	37 H1		Keswick Ct	36 D3		Reddington Dri	37 H6		Winvale	36 B5
Court Cres	36 B1		King Edward St	36 B5		Redwood Gdns	36 A2		Winwood	37 F2
Cranbourne Rd	36 A4		Kings Rd	36 B6		Regent Ct	36 C2		Withy Croft	37 H2
Cromwell Dri	36 C1		Kirkwall Spur	36 B1		Richmond Cres	36 D4		Woodland Av	36 B3
Crown Corner	36 C4		Knolton Way	37 E1		Rixon Clo	37 H1			
						Rochfords Gdns	37 F3		Yew Tree Rd	36 D6
Dandridge Clo	37 G6		La Roche Clo	37 F5		Ronaldsay Spur	36 B1		York Av	36 A2
Darvills La	36 B5		Ladbrooke Rd	36 A5		Rossiter Clo	37 H6			
						Rutland Av	36 A1			

THATCHAM

Agricola Way	39 F5
Alexander Rd	39 F5
Alston Ms	38 D5
Appelford Clo	39 F5
Arkle Way	38 A4
Arrowsmith Way	39 G5
Ashbourne Way	38 C4
Ashman Rd	39 G5
Ashmore Green Rd	38 C1
Ashworth Dri	38 D5
Baily Av	38 C3
Barfield Rd	38 C3
Barley Clo	39 G5
Barley Ct	39 G5
Bath Rd	38 B3
Bath Rd	39 F4
Beancroft Rd	39 E5
Beech Walk	39 F5
Benham Hill	38 A3
Berkshire Dri	39 H5
Betteridge Rd	39 G5
Beverley Clo	38 D2
Billington Way	38 D2
Blackdown Way	38 D5
Bluecoats	39 E3
Blythe Rd	39 F5
Bodmin Clo	38 D5
Bollingbroke Way	39 G4
Boscowen Way	39 G5
Botany Clo	39 G4
Bourne Arch	38 C3
Bourne Rd	38 C3
Bowes Rd	39 E5
Bowling Green Rd	38 B2
Braemore Clo	39 E5
Bramwell Clo	39 G5
Brent Clo	39 E5
Broadway	39 E4
Brooks Rd	39 F3
Brownfield Rd	38 D4
Browning Clo	38 D3
Buchanan Sq	39 F6
Burns Walk	38 D3
Cairngorm Rd	39 E5
Callard Dri	38 C3
Chapel Ct	39 F4
Chapel St	39 E4
Chapman Walk	38 C3
Chesterton Rd	38 D2
Cholsey Rd	39 G4
Church Gate	39 E4
Church La	39 E4
Cochrane Clo	39 F4
Cold Ash Hill	38 D2
Coniston Clo	38 B4
Conway Dri	38 C2
Coombe Ct	39 F4
Coopers Cres	38 D3
Corderoy Clo	39 G5
Cropper Clo	39 H4
Crowfield Dri	38 C4
Crown Acre Clo	38 D4
Crown Mead	38 D4
Curlew Clo	38 D4
Cygnet Clo	38 C4
Danvers Clo	39 E5
Dart Clo	38 C2
Denton Clo	38 D5
Derwent Rd	38 B4
Domoney Clo	39 F4
Doublet Clo	38 B4
Draper Clo	39 E5
Druce Way	39 E4
Dryden Clo	39 E2
Dunstan Rd	39 F3
Edwin Clo	39 G4
Eliot Clo	38 D2
Elm Av	39 F4
Elm Gro	38 C2
Elmhurst Rd	38 C2
Ennerdale Way	38 B4
Enterprise Way	39 H5
Ermine Walk	38 C4
Evreux Clo	39 G5
Exmoor Rd	38 D4
Falmouth Way	39 G4
Ferndale Ct	39 E4
Flag Staff Sq	39 F5
Flecker Clo	39 E2
Fokerham Dri	39 G5

Foxhunter Way	38 B4
Fromont Dri	39 E4
Fuller Clo	39 G5
Fyfield Clo	39 E5
Fylingdales	38 D5
Glaisdale	38 D5
Glenmore Clo	39 E6
Golding Clo	39 G4
Goldsmith Clo	38 D2
Goose Green Way	39 F4
Gordon Rd	38 B2
Grassington Pl	39 E5
Grassmead	39 G5
Great Barn Ct	39 E4
Green La	38 D4
Griffiths Clo	39 F5
Grindle Clo	38 D2
Hambridge La	38 A5
Hammond Clo	39 G5
Hardy Clo	38 D2
Hartley Way	39 F3
Hartmead Rd	39 F4
Harts Hill Rd	39 F3
Hatchgate Clo	38 D1
Hazel Gro	39 E2
Heardman Clo	39 G5
Heath La	38 D2
Hebden Clo	39 E6
Henwick La	38 B2
Heron Way	38 C4
High St	39 E4
Holywell Ct	39 E4
Horne Rd	39 E5
Humber Clo	38 C2
Hurford Dri	39 G4
Ilkley Way	38 D5
Jedburgh Clo	39 G4
John Hunt Clo	39 F5
Justice Clo	39 G5
Keighley Clo	38 D5
Kendal Clo	38 D3
Kennet Clo	39 F4
Kestrel Clo	38 C4
Kipling Clo	38 D2
Lamb Clo	38 D2
Lancaster Clo	38 D3
Lawrences La	39 E2
Link Way	38 C3
Longbridge Rd	39 F5
Longcroft Rd	39 F5
Loundyes Clo	38 C3
Lower Way	38 A4
Lyon Clo	39 G5
Magpie Clo	38 D4
Malham Rd	38 D4
Malthouse Clo	39 F5
March Rd	39 F3
Masefield Rd	39 E3
Maynard Clo	39 E2
Mayow Clo	39 G5
Meadow Clo	38 D4
Medway Clo	38 C2
Mersey Way	38 C2
Montacute Dri	39 G5
Mount Rd	39 E3
Munkle Marsh	39 G4
Neville Dri	39 F4
Newbolt Clo	38 D2
Nideggen Clo	39 E4
Norlands	38 D2
Northfield Rd	38 D3
Northway	38 D2
Oak Tree Rd	39 F5
Park Av	39 E3
Park La	39 E2
Parkside Rd	39 E2
Pavy Clo	39 G5
Paynesdown Rd	38 C4
Peachey Clo	39 G5
Pegasus Clo	38 B3
Pentland Clo	39 E5
Pipers La	39 G5
Pipers Way	39 G5
Pipit Clo	38 C4
Poffley Pl	39 G5
Porlock Clo	39 E5

Pound La	38 B4
Pr Hold Rd	38 B4
Quarrington Clo	39 F5
Robertsfield	38 A4
Roman Way	38 C3
Rope Walk	38 D4
Rosedale Gdns	38 D5
Rosier Clo	39 G5
Rudland Clo	39 E5
Rydal Dri	38 B4
Sagecroft Rd	38 D3
St Johns Rd	38 D4
St Marks Clo	38 D4
Sargood Clo	39 F5
Scrivens Mead	39 G4
Severn Clo	38 C2
Shakespeare Rd	38 D3
Shelley Rd	38 D3
Skillman Dri	39 G4
Snowdon Clo	39 E5
Somerton Gro	38 D5
Southdown Rd	38 A4
Southend	38 D1
Spackman Clo	39 E5
Spriggs Clo	39 E5
Spurcroft Rd	39 E5
Station Rd	39 E4
Stephenson Clo	39 E3
Stoney La	39 F4
Swansdown Clo	38 C4
Swansdown Walk	38 C4
Sydney Clo	39 F4
Tadham Pl	38 D5
Tennyson Rd	38 D3
Thames Rd	38 C2
The Alders	39 E3
The Close	38 C3
The Firs	38 D3
The Frances	39 E3
The Grove	39 E3
The Haywards	39 E3
The Hollands	39 F4
The Martins	39 G5
The Moors	39 E4
The Quantocks	39 E5
The Thackerays	39 E5
The Turnery	38 D4
The Waverleys	39 E3
Thompson Clo	39 F5
Tomlin Clo	39 F5
Trent Cres	38 C2
Turners Rise	39 F5
Turnfields Rd	39 E4
Turnpike Rd	38 A3
Tyne Way	38 C2
Ullswater Clo	38 C4
Victor Rd	39 F4
Vincent Rd	39 F3
Walsingham Way	39 E5
Webbs Acre	39 G5
Wenlock Way	39 E5
Westerdale	39 E5
Westfield Cres	38 C3
Westfield Rd	38 C2
Westland	38 C3
Wheelers Grn Way	39 F5
Whitelands Rd	39 E3
Wilfred Way	39 G5
William Clo	39 E5
Windermere Way	38 C4
Winston Way	38 B4
Wordsworth Rd	38 D3

TWYFORD

Amberley Dri	40 B2
Arnside Clo	40 C2
Badger Dri	40 C2
Bolwell Clo	40 D5
Bridge Farm La	40 A2
Broad Hinton	40 D5
Broadwater Rd	40 C6
Brook St	40 B4
Byron Clo	40 C4
Byron Rd	40 C5
Carlisle Gdns	40 B2
Castle End Rd	40 D3

Chapel Row	40 B4
Chaseside Av	40 B2
Cheriton Av	40 C2
Church La	40 D3
Church St	40 B4
Colleton Dri	40 C5
Crest Clo	40 C2
Harrison Clo	40 D5
Hermitage Dri	40 B4
Heron Clo	40 C2
High St	40 B4
Hilltop Rd	40 C2
Hurst Park Rd	40 C6
Hurst Rd	40 C5
Kibblewhite Cres	40 C3
Kingfisher Ct	40 D5
Lincoln Gdns	40 B3
Loddon Hall Rd	40 C3
London Rd	40 B4
Longfield Rd	40 B3
Mallard Clo	40 C6
Malvern Way	40 B2
Maple Bank	40 C3
Middlefields	40 C2
Milton Way	40 D3
New Bath Rd	40 A3
New Rd	40 B2
New Rd, Ruscombe	40 D3
Northbury Av	40 C3
Northbury La	40 C3
Old Bath Rd	40 C3
Orchard Est	40 C4
Paddock Heights	40 C5
Park La	40 A6
Pennfields	40 C2
Pine Gro	40 B4
Polehampton Clo	40 B4
Ruscombe La	40 D3
Ruscombe Pk	40 C3
Ruscombe Rd	40 C3
Ruscombe Rd Ind Est	47 C4
St James Clo	40 C3
St Michaels Ct	40 C2
St Swithins Ct	40 B4
South View Clo	40 C2
Southbury La	40 D4
Springfield Pk	40 C4
Stanlake Rd	40 D5
Station Rd	40 B4
Swans Ct	40 C5
Sycamore Dri	40 C3
The Grove	40 C4
Treacher Ct	40 C4
Troutbeck Clo	40 B2
Verey Clo	40 D5
Wagtail Clo	40 C4
Walnut Tree Clo	40 D3
Waltham Rd	40 C4
Wargrave Rd	40 B1
Wensley Clo	40 C3
Wessex Gdns	40 C5
Westview Rd	40 C2
Willow Dri	40 B2
Winchcombe Rd	40 C5
Woodpecker Clo	40 C5
Yewhurst Clo	40 B2

UXBRIDGE

Alder Rd	41 A1
Alexandra Rd	41 B6
Atwells Yd	41 B3
Austin Waye	41 A5
Bakers Rd	41 B3
Bakers Yd	41 B3
Barnfield Pl	41 A4
Bassett Rd	41 A3
Bawtree Rd	41 B2
Belmont Clo	41 C2
Belmont Rd	41 B3
Bettles Clo	41 A5
Blackmore Way	41 B2

Bonsey's Yd 41 C3
Braybourne Clo 41 A2
Brearley Clo 41 D1
Bridge Rd 41 A6
Burness Clo 41 C6

Cambridge Rd 41 B1
Caxton Dri 41 B6
Chapel St 41 B4
Charter Pl 41 B4
Chequers Sq 41 B3
Chiltern View Rd 41 B6
Chippendale All 41 C4
Chippendale Waye 41 C3
Cobden Clo 41 A4
Concorde Clo 41 D6
Cornfield Clo 41 C6
Cornwall Rd 41 C1
Cotswold Clo 41 B5
Cowley Mill Rd 41 A6
Cowley Rd 41 B6
Cricketfield Rd 41 B4
Cross St 41 B4
Crown Walk 41 B3
Cumbrian Way 41 C4

Dawes Rd 41 D6
Derby Rd 41 B5

Elm Lawn Clo 41 D2

Fairfield Rd 41 B1
Fassnidge Vw 41 A3
Frays Waye 41 A4

Gatting Way 41 D1
George St 41 B3
Glebe Rd 41 A6
Grainges Yd 41 B3
Greys Rd 41 D4
Grove Rd 41 C3
Grove Waye 41 C3

Harefield Rd 41 B2
Hawthorn Rd 41 A1
Heron Clo 41 C1
High St 41 B3
Hillingdon Rd 41 B4
Hinton Rd 41 B4
Honeycroft Hill 41 D3
Hows Clo 41 A4
Hows Rd 41 A4

Iffley Clo 41 C3

Jackson Clo 41 D3
Jackson Rd 41 D3

Lancaster Rd 41 B2
Lancresse Clo 41 B2
Lawn Rd 41 A4

Manor Waye 41 C5
Market Sq 41 B3
Martin Clo 41 D6
Maylands Dri 41 C1
Mead Rd 41 C2
Medman Clo 41 B6
Mercer Walk 41 B3
Merry Fields 41 C6
Mill Av 41 A6
Montague Rd 41 C3
Myddleton Rd 41 B5

New Winsdor St 41 A4
Norfolk Rd 41 C2
Nursery Waye 41 C5

Orchard Waye 41 C6
Osborn Rd 41 A3
Oxford Rd 41 A1

Pantile Walk 41 B3
Park Rd 41 C4
Park Rd E 41 C6
Penrith Clo 41 C4
Press Rd 41 B2

Redfern Clo 41 A4
Redford Way 41 B3
Rockingham Clo 41 A4
Rockingham Rd 41 A4
Rushes Mead 41 A4

St Andrews Rd 41 D4
St Lukes Rd 41 D4

South Common Rd 41 D1
Stanley Clo 41 B5

Tachbrook Rd 41 A6
The Close 41 D3
The Greenway 41 B6
The Hermitage 41 C2
The Lynch 41 A3
Thompson Rd 41 D4
Trumper Way 41 A4
Turnpike 41 D6

Valley Rd 41 D6
Victoria Rd 41 A3
Vine St 41 B4

Walford Rd 41 B6
Wallace Clo 41 D6
Warwick Pl 41 B2
Waterloo Rd 41 A4
Wellington Clo 41 A4
West Common Rd 41 C1
Westcott Waye 41 A5
Whitehall Clo 41 B4
Whitehall Rd 41 B4
Willow Way 41 A1
Wilmar Rd 41 B2
Windsor St 41 B4
Wyvern Way 41 A4

York Rd 41 C3

WARGRAVE

Autumn Walk 42 B4

Backsideans 42 B4
Bayliss Rd 42 B4
Beverley Gdns 42 C4
Blakes Rd 42 C3
Braybrooke Gdns 42 B4
Braybrooke Rd 42 B4

Church St 42 B4
Clifton Rise 42 C4

Dark La 42 C3
Dunnock Way 42 D3

East View Rd 42 D3
Elizabeth Ct 42 C4
Emma La 42 C4

Ferry La 42 B4
Fidlers Walk 42 C4

Hamilton Rd 42 C4
High St 42 B4
Highfield Pk 42 D3
Hill Lands 42 C3

Langham Way 42 C3
Loddon Dri 42 A4

McCraes Wk 42 C4
Mill La 42 A3
Mumbery Hill 42 C4

Newalls Rise 42 C3

Purfield Dri 42 C4

Recreation Rd 42 C3
Ridge Way 42 C3
Ryecroft Rd 42 C3

School Hill 42 C4
School La 42 B4
Silverdale Rd 42 C4
Station Rd 42 A4

The Bothy 42 B3
The Copse 42 C3
The Spur 42 C3
The Vinery 42 B3
The Walled Gdn 42 C3

Victoria Rd 42 C4

Wargrave Hill 42 B3
Watermans Wy 42 A4
Willow La 42 B4

WICK HILL

Arnett Clo 43 B3

Bankside 43 C4
Barkham Ride 43 A2
Billing Av 43 B4
Birch Rd 43 C3
Briarwood 43 A3
Buchanan Dri 43 B3
Burchett Coppice 43 B2
Burnt Oak 43 C2

Cambrian Way 43 D2
Carolina Pl 43 A3
Challenor Clo 43 B3
Charlton Clo 43 B2
Chivers Dri 43 B3
Church Hams 43 A4
Church La 43 B6
Columbia Ct 43 A3
Copse Way 43 A3
Croft Rd 43 C1
Cypress Clo 43 D2

Dare Clo 43 B4
Ditchfield La 43 B2
Dodsells Well 43 B2
Drake Clo 43 B3

Elgarth Dri 43 B2

Fernbank 43 A3
Finchampstead Rd 43 B5
Fir Cottage Rd 43 B2
Firs Clo 43 B4
Foxcote 43 C4

Garret Rd 43 B2
Garston Gro 43 B1
Gibbs Clo 43 B4
Gorse Ride Nth 43 B3
Gorse Ride Sth 43 B4

Hazelbank 43 A4
Heath Ride 43 C4
Heather Clo 43 A3

Ingleden 43 C3

Jerrymoor Hill 43 C3
Jubilee Rd 43 B6

Kelsey Av 43 B4
Kiln Ride 43 C2

McCarthey Way 43 B2
Manor Park Dri 43 A3
Maryland 43 A3
Moor Clo 43 A3
Mornington Av 43 B2

Nashgrove La 43 B1
Ninemile Ride 43 A4

Orbit Clo 43 B4
Oregon Walk 43 A3

Pine Dri 43 C3

Radical Ride 43 B3
Range Rd 43 D3
Roycroft La 43 B2
Russley Grn 43 C1

St James Rd 43 A2
Sandhurst Rd 43 D2
Shenstone Clo 43 C2
Springdale 43 B2
Summit Clo 43 C4

Tanglewood 43 C3
The Lea 43 B2
The Spinney 43 B2
Thomas Lea 43 B2
Ticknor Dri 43 C3
Tintagel Rd 43 D2
Tomlinson Dri 43 D3

Vermont Woods 43 A3

Warren Clo 43 B4
Warren La 43 A5
Washington Gdns 43 C3
Watkins Clo 43 B3
Watson Clo 43 B1
Waverley Way 43 B1

White Horse La 43 A5
Whittle Clo 43 B4
Wick Hill La 43 C4
Wild Briar 43 C3
Wildcroft Dri 43 C1
Willow Dale 43 B3
Wimbushes 43 A4
Windsor Ride 43 C3
Woodside Clo 43 A2

WINDSOR & ETON

Abbotts Walk 46 C4
Addington Clo 47 E5
Adelaide Sq 47 H4
Albany St 47 G4
Albert Rd 47 H5
Albert St 47 F3
Alden Vw 46 B3
Alexandra Rd 47 G4
Alider Mans 46 B3
Alma Rd 47 G4
Arthur Rd 47 F3
Ash La 46 B4
Aston Mead 46 C3
Atherton Ct 47 G2
Athlone Sq 47 G3

Bachelors Acre 47 H3
Bailey Clo 47 E4
Baldwins Shore 47 G1
Ballard Grn 46 C2
Balmoral Gdns 47 G5
Barry Av 47 G2
Basford Way 46 B5
Beaumont Rd 47 G4
Bedford Rd 46 B3
Bell Par 46 D4
Bell Vw 46 D5
Bell View Clo 46 B4
Benning Clo 46 B5
Bexley St 47 F3
Birch Gro 46 B3
Birchington Rd 47 E4
Black Horse Clo 46 B4
Bolton Av 47 G5
Bolton Cres 47 G5
Bolton Rd 47 G5
Bone Av 47 G5
Bourne Av 47 G3
Bowes-Lyon Clo 47 G3
Bradshaw Clo 46 C3
Bridgeman Dri 47 E4
Bridgewater Ter 47 G3
Brocas St 47 G2
Brook St 47 H4
Brudenell 46 D5
Bryer Pl 46 B5
Buckland Cres 46 D3
Bulkeley Av 47 F4
Burnetts Rd 46 C3
Burnham Clo 46 B4
Burton Way 46 C5
Butlers Clo 46 B3
Byron Ct 47 E5

Carey Clo 47 F5
Carter Clo 47 E4
Castle Hill 47 H3
Cavalry Cres 47 F5
Cawcott Dri 46 C3
Chantry Clo 47 E3
Charles St 47 G3
Charlton 46 A4
Chestnut Dri 46 D6
Christian Sq 47 G3
Church Clo 47 G1
Church La 47 H3
Church Ter 46 C4
Claremont Rd 47 G3
Clarence Cres 47 G3
Clarence Rd 47 F3
Clewer Av 47 E4
Clewer Ct 47 E2
Clewer New Town 47 E4
Clewer Pk 47 E2
Clewer Rd, Clewer 47 E2
Clewer Rd, Windsor 46 C4
Clifton Rise 46 B3
Cockpits Clo 46 C3
College Cres 47 F4
Convent Rd 46 D4
Coombemere Clo 47 F4
Copper Beech Clo 46 B3
Cranbourne Av 46 D4
Cross Oak 47 E4

Name	Ref
Dagmar Rd	47 G4
Datchet St	47 H2
Dawson Clo	47 E4
Dean Clo	46 B5
Dedworth Dri	46 D3
Dedworth Rd	46 A4
Devereux Rd	47 G4
Dorset Rd	47 G4
Dower Pk	46 C6
Duke St	47 F2
Duncannon Cres	46 B5
Duncroft	46 D5
Dyson Clo	47 F5
East Cres	46 D3
Edinburgh Gdns	47 H5
Ellison Dri	46 D5
Elm Rd	47 F5
Errington Dri	47 E3
Eton Ct	47 G2
Eton Sq	47 H2
Etonwick Rd	47 H1
Fairacres	46 B4
Fairlawn Pk	46 C6
Fairlight Av	47 H4
Farm Yd	47 H2
Fawcett Rd	47 F3
Filmer Rd	46 B4
Firs Av	46 D5
Forest	46 C4
Foster Av	46 C5
Fountain Gdns	47 H5
Frances Rd	47 G5
Frances St	47 G4
Franklyn Cres	46 B5
Frymley Way	46 B3
Furness	46 A4
Gallys Rd	46 B3
Gilman Cres	46 B5
Gloucester Pl	47 H4
Gordon Rd	46 D5
Goslar Way	47 F4
Goswell Hill	47 G3
Goswell Rd	47 G3
Gratton Dri	46 C6
Green Acre	46 C4
Green La	47 E4
Grove Rd	47 G4
Guards Rd	46 B4
Gwynne Clo	46 C3
Hanley Clo	46 B3
Hanover Way	46 D3
Harcourt Rd	46 C3
Harrington Clo	46 D6
Haslemere Rd	47 E3
Hatch La	46 D5
Hawtrey Rd	47 G4
Hayse Hill	46 B3
Helena Rd	47 H4
Helston La	47 E3
Hemwood Rd	46 B5
Hermitage La	47 E5
Highfield	47 H3
Hilltop Rd	46 C5
Holly Cres	46 B4
Homers Rd	46 B3
Huddlesway	46 B3
Hylle Clo	46 C3
Illingworth	46 C5
Imperial Rd	47 E5
Jacobs Clo	46 C3
James St	47 G3
Keats La	47 G1
Keeler Clo	46 C5
Keepers Farm Clo	46 C4
Kenneally	46 A4
Kentons La	46 C4
Keppel St	47 H4
Kimber Clo	46 D5
King Edwart Ct	47 G3
King Edward VII Av	47 H2
Kings Field	46 B3
Kings Rd	47 H4
Kingstable St	47 H2
Knights Clo	46 B3
Lammas Ct	47 G4
Leigh Sq	46 B4
Liddell Pl	46 B5
Little Buntings	46 D5
Lodge Way	46 C5
Longmead	46 C3
Loring Rd	46 D3
Losfield Rd	46 C3
Lovejoy La	46 B4
Luff Clo	46 C5
Lyell	46 A5
Madeira Walk	47 H3
Maidenhead Rd	46 A3
Manor Farm Clo	46 D5
Manor Rd	46 C4
Mansell Clo	46 C3
Marbeck Clo	46 B3
Martins Clo	46 B3
Meadow La	47 F1
Merwin Way	46 B4
Mill La	47 E2
Monks Rd	46 B4
Montpelier	47 G4
Mountbatten Sq	47 G3
Needham Clo	46 C3
Nelson Rd	46 D5
Newbury Cres	46 B4
Nicholls	46 A5
North Clo	46 D3
Oak La	47 E3
Oakley Grn Rd	46 A3
Orchard Av	47 E3
Orwell Clo	47 G5
Osborne Ct	47 G4
Osborne Mews	47 G4
Osborne Rd	47 G4
Oxford Rd	47 F3
Oxford St	47 G3
Park Clo	47 H4
Park Corner	46 C5
Park St	47 H3
Parsonage La	47 E3
Peascod St	47 G3
Peel Clo	47 E5
Perrycroft	46 C5
Pierson Rd	46 B3
Poolmans Rd	46 B5
Prince Consort Cotts	47 G4
Princess Av	47 F5
Priors Rd	46 B5
Queen Annes Rd	47 G6
Queens Acre	47 G6
Queens Rd	47 G4
Rays Av	46 D2
Rectory Clo	47 E3
Regent Ct	47 H4
River St	47 G2
Roses La	46 C4
Rowland Clo	46 B5
Ruddlesway	46 B3
Russel St	47 G3
Rydings	46 C5
Ryecroft	46 D5
St Albans St	47 H3
St Andrews Av	46 D4
St Andrews Cres	46 D4
St Georges Clo	46 C3
St Johns Dri	46 D4
St Johns Rd	47 E4
St Leonards Av	47 G4
St Leonards Hill	46 C5
St Leonards Rd	47 G4
St Marks Pl	47 G4
St Marks Rd	47 G4
Sawyers Clo	46 C2
Sheepcote Rd	46 C4
Sheet St	47 H4
Sherbourne Dri	46 D6
Shirley Av	46 D3
Shirley Hall Clo	46 B3
Sidney Rd	46 A5
Smiths La	46 C4
Snowdon Clo	46 B6
South Meadow La	47 G2
Spinners Walk	47 G3
Springfield Clo	47 F4
Springfield Rd	47 F4
Stephenson Dri	47 E2
Stirling Clo	46 B4
Stovell Rd	47 F2
Stroud Clo	46 B5
Stuart Clo	46 D4
Stuart Way	46 C4
Sunbury Rd	47 H1
Tangier Ct	47 H1
Tangier La	47 H1
Tarbay La	46 A4
Temple Rd	47 G4
Testwood Rd	46 B3
Thames Av	47 H2
Thames Mead	46 C3
Thames Side	47 H2
Thames St	47 H2
The Hatch	46 A2
The Parade	46 B3
Tinkers La	46 B4
Trinity Pl	47 G4
Tudor Way	46 C3
Turnoak Park	46 C5
Tyrell Gdns	46 D5
Upcroft	47 E5
Vale Rd	46 D2
Vansittart Rd	47 F3
Victor Rd	47 G5
Victoria St	47 G3
Washington Dri	46 C5
Wells Clo	47 E3
West Cres	46 D3
Westmead	47 F5
White Horse Rd	46 B5
Whiteley Clo	46 C2
William St	47 G3
Wilton Cres	46 B5
Windmill Clo	47 F4
Winkfield Rd	46 D6
Witney Clo	46 C3
Wolf La	46 B5
Wood Clo	47 G6
Woodland Av	46 D6
Wright	46 B5
Wyatt Clo	46 B5
York Av	47 F4
York Rd	47 F4

WINNERSH

Name	Ref
Albany Park Dri	44 A2
Allnatt Av	44 B4
Annesley Gdns	44 C2
Arbor La	44 A2
Arun Clo	44 B4
Ashton Rd	44 E4
Astor Clo	44 C2
Baslow Rd	44 A3
Bathurst Rd	44 A2
Bayley Ct	44 B4
Bearwood Path	44 A2
Beckford Clo	44 E4
Birchmead	44 D3
Bluebell Meadow	44 B2
Borrowdale Rd	44 A1
Bredon Rd	44 E4
Brimblecombe Clo	44 F4
Chatsworth Av	44 A3
Church La	44 B3
Churchill Dri	44 A3
Clarendon Clo	44 C3
Clifton Rd	44 F4
Commons Rd	44 E4
Cornfield Grn	44 E4
Danywern Dri	44 B3
Davis Clo	44 B3
Davis St	44 C1
Davis Way	44 C1
Deerhurst Av	44 C3
Defford Clo	44 E4
Delane Dri	44 A3
Donnington Pl	44 C3
Douglas Grange	44 C1
Dunt Av	44 E1
Dunt La	44 D1
Eastbury Pk	44 D3
Eden Way	44 A4
Elmley Clo	44 E4
Emmbrook Gate	44 F4
Emmbrook Rd	44 E4
Emmbrook Vale	44 F4
Eskdale Rd	44 A1
Field Way	44 C3
Forest Rd	44 D4
Fulbrook Clo	44 F4
Garth Clo	44 B3
Goddard Ct	44 B3
Grassmere Clo	44 B4
Green La	44 C4
Green La	44 D1
Greenwood	44 C2
Grovelands Av	44 C3
Grovelands Clo	44 C2
Harefield Clo	44 C2
Harman Ct	44 B3
Isis Clo	44 B4
Kelburne Clo	44 B2
King Street La	44 A4
Laburnum Rd	44 B4
Lenham Clo	44 D4
Little Hill Rd	44 C1
London Rd	44 B4
Lowther Clo	44 E4
Lowther Rd	44 E4
Maidensfield	44 C3
Maple Clo	44 C2
Matthewsgreen Rd	44 F4
May Fields	44 A4
Meadow Vw	44 C2
Melbourne Av	44 B4
Melody Clo	44 B2
Merryhill Chase	44 C2
Merryhill Green La	44 C2
Mole Rd	44 A4
Mungells La	44 B2
Overbury Av	44 E4
Pheasant Clo	44 B3
Poplar La	44 C2
Primrose La	44 C2
Priory Ct	44 C2
Reading Rd	44 A2
Reynards Clo	44 C3
Robin Hood Way	44 C2
Robinhood La	44 B3
Roundabout La	44 D4
Russell Way	44 A4
Sadlers La	44 C4
St Catherines Clo	44 A4
St Marys Rd	44 A4
Sandstone Clo	44 B4
Sherwood Rd	44 C2
Snowdrop Gro	44 B2
Summerfield Clo	44 F3
Targett Ct	44 B3
The Orchard	44 A2
The Priory	44 C2
Toutley Clo	44 E4
Toutley Rd	44 F3
Turnstone Clo	44 A2
Watmore La	44 C2
Wedderburn Clo	44 C3
Welby Cres	44 A3
Westfield Rd	44 A3
Wharfdale Rd	44 A1
Wilmott Clo	44 B3
Wilson Ct	44 B3
Windermere Clo	44 C2
Winnersh Gate	44 C3
Winnersh Gro	44 C4
Woodhurst	44 F4

WOKINGHAM

Name	Ref
Abbey Clo	49 E2
Acorn Dri	49 E2
Agate Clo	48 A2
Agincourt Clo	48 B3
Albert Rd	48 D4
Alderman Willey Clo	48 D3
All Saints Clo	49 E1
Amethyst Clo	48 A2
Andrew Clo	49 G4
Antares Clo	48 B3
Apple Clo	48 B4
Aquila Clo	48 B3
Arthur Rd	48 C3

Achridge Rd 49 F1
Ashville Way 48 D4
Astley Clo 48 B2

Barkham Rd 48 A6
Barkhart Dri 49 E2
Barkhart Gdns 49 E2
Barrett Cres 49 F3
Battys Barn Clo 49 F4
Beale Clo 49 E2
Bean Oak Rd 49 G3
Beckett Clo 49 G3
Bedford Gdns 48 B2
Bedfordshire Way 48 A3
Beechnut Clo 48 B4
Bell Foundry La 49 E1
Benning Rd 49 F1
Beryl Clo 48 A2
Binfield Rd 49 G3
Bird Mews 49 E3
Bishops Dri 49 E2
Blagrove Dri 48 C5
Blagrove La 48 C5
Blake Clo 49 G1
Blenheim Clo 48 B3
Bowyer Cres 49 F1
Bridge Clo 48 B2
Broad St 49 E3
Brook Clo 48 C1
Brookside 48 C2
Broom Gro 48 A5
Buckhurst Gro 49 H4
Buckthorn Clo 49 G2
Budges Gdns 49 G2
Budges Rd 49 F1

Cambridgeshire Clo 48 A3
Camelia Way 48 A2
Campion Way 49 G2
Cantley Cres 48 C1
Carey Rd 49 E4
Caroline Dri 48 C2
Cedar Clo 49 E2
Chaffinch Clo 48 B4
Charwood Rd 49 G3
Chaucer Clo 49 G3
Chaucer Way 48 A4
Cheeseman Clo 49 F2
Chestnut Av 48 A3
Chetwode Clo 49 G3
Clare Av 49 E2
Clay La 49 H3
Clifton Rd 48 C1
Cockpit Path 49 E4
Copse Dri 48 C2
Cornflower Clo 48 A2
Cornwall Clo 48 A3
Coronation Sq 49 F2
Crail Clo 48 D6
Crecy Clo 48 A3
Crescent Rd 49 E4
Crocus Clo 48 A2
Cross St 49 F3
Crutchley Rd 49 F2
Culloden Way 48 A3
Cumberland Way 48 A3
Curl Way 48 C4

De Vitre Grn 49 H2
Deacon Clo 49 E1
Dean Gro 49 E2
Denmark St 49 E4
Denton Rd 49 E3
Derwent Clo 48 B3
Devon Clo 48 A3
Diamond Way 48 A2
Dieppe Clo 48 B3
Doles Hill 48 A6
Doles La 48 A5
Dorset Way 48 A3
Duncan Dri 49 G4
Dunkirk Clo 48 A3
Durham Clo 48 A3

Eagle Clo 48 B4
Easthampstead Rd 49 F3
Eastheath Av 48 D5
Eastheath Gdns 48 D5
Elizabeth Rd 49 F3
Ellison Way 49 E3

Elms Rd 49 E4
Emerald Clo 48 A2
Emmbrook Gate 48 C1
Emmbrook Rd 48 C1
Emmview Clo 48 C2
Eustace Cres 49 F1
Evendons Clo 48 D6
Evendons La 48 C6
Evergreen Way 48 C4

Fairview Rd 49 E4
Falcon Way 48 B3
Finchampstead Rd 48 D6
Fishponds Clo 48 C5
Fishponds Rd 48 C5
Flamingo Clo 48 B4
Fleet Clo 48 B3
Forest Rd 48 A1
Foxglove Clo 48 A2
Frederick Pl 48 D3
Freesia Clo 48 A2
Froghall Dri 49 G3

Gipsy La 49 E4
Glebelands Rd 49 E2
Goodchild Rd 49 F3
Goodrings Grn 49 H3
Gorrick Sq 48 D6
Green Croft 49 G1
Green Dri 49 G4
Gull Clo 48 B4

Hampshire Way 48 A3
Harmar Clo 49 G3
Hart Dyke Clo 48 D6
Havelock Rd 48 D3
Hawkes Clo 48 C2
Hawkins Way 49 G3
Heath Clo 48 D2
Heathlands Rd 49 G6
Heddington Clo 49 F1
Heddington Dri 49 F1
Heelas Rd 48 C3
Heron Rd 48 B3
Herons Way 49 G2
Hertford Clo 48 A4
Highfield Clo 49 E3
Holmes Clo 48 C5
Holmes Cres 48 C5
Holt La 48 D2
Howard Rd 49 E4
Hughes Rd 49 G2
Humber Clo 48 B2
Huntingdonshire Clo 48 A3
Hutsons Clo 49 F1

Japonica Clo 48 A5
Jasmine Clo 48 A2
Jubilee Av 48 D2
Jupiter Way 48 B3
Jutland Clo 48 B3

Keephatch Rd 49 G1
Kendrick Clo 49 E4
Kennet Ct 48 B3
Kent Clo 48 A3
Kesteven Way 48 A3
Kestrel Way 48 B3

Lalande Clo 48 B3
Landen Ct 48 D5
Langborough Rd 49 E4
Larch Av 48 D2
Larkspur Clo 48 A2
Latimer Rd 48 D3
Laud Way 49 G3
Laurel Clo 48 B4
Lawrence Clo 49 F3
Lenny Clo 49 F1
Lime Clo 48 C4
Limmerhill Rd 48 B4
Linden Clo 48 B5
Lindsey Clo 48 A3
London Rd 49 F3
Longs Way 49 G2
Lowther Clo 48 B1
Lowther Rd 48 B1
Luckley Path 49 E4
Luckley Rd 48 D6
Luckley Wood 49 D6

Magnolia Way 48 B4
Mansfield Rd 48 C4
Market Pl 49 E3
Marks Rd 48 C1
Mars Clo 48 B3
Martins Dri 48 D1
Matthewsgreen Rd 48 C1
Mays Rd 49 G3
Meadow Rd 48 C3
Meadow Walk 48 C3
Meadow Way 48 C3
Medina Clo 48 B2
Medway Clo 48 B2
Mercury Av 48 B3
Mill Clo 48 C2
Millmead 48 C2
Milton Dri 48 D2
Milton Gdns 48 D3
Milton Rd 48 D2
Minden Clo 48 A3
Moles Clo 49 F4
Molly Millars Bri 48 D5
Molly Millars Clo 48 D5
Molly Millars La 48 C4
Monkswood Clo 49 G2
Mons Clo 48 A3
Moores Grn 49 G1
Mount Pleasant 48 D3
Mower Clo 49 H2
Murdoch Rd 49 F4
Murray Rd 48 C3
Mylne Sq 49 G3

Neptune Clo 48 B3
Norfolk Clo 48 A3
Norreys Av 49 F3
North Way 48 A2
Norton Rd 49 E4

Oaklands Dri 48 C5
Old Woosehill La 48 B2
Opal Way 48 A2
Orchard Clo 49 F3
Orchard Pl 49 E3
Ormonde Rd 48 D4
Osborne Rd 49 E3
Osterley Clo 49 H4
Owl Clo 48 B4
Oxford Rd 48 C2

Pages Croft 49 F4
Paice Grn 49 F2
Palmer School Rd 49 E3
Park Av 48 D3
Park Rd 48 D3
Patten Ash Dri 49 G2
Peach St 49 E3
Peregrine Clo 48 B4
Perkins Way 48 C4
Petrel Clo 48 B4
Phoenix Clo 48 B3
Pigott Rd 49 F1
Plough La 49 H2
Plover Clo 48 B4
Priest Av 49 H3
Proctor Rd 49 H3

Quartz Clo 48 A2

Rances La 49 G3
Reading Rd 48 A1
Rectory Clo 49 E3
Rectory Rd 49 E3
Reeves Way 48 C5
Rembrandt Clo 48 A3
Richmond Rise 48 A2
Riding Way 48 A3
Rook Clo 48 B4
Rose St 49 E3
Rotherfield Av 48 C2
Roundabout La 48 A1
Rowan Clo 48 B4
Ruby Clo 48 A2
Ruskin Way 48 A3

Sadlers La 48 A1
St Heliers Clo 48 D6
St Pauls Gate 48 C3
Sandy La 48 A5
Sapphire Clo 48 A2

Sarum Cres 49 F2
Saturn Clo 48 B3
School Rd 49 F3
Scots Dri 48 A1
Seaford Rd 49 F3
Sewell Av 48 C1
Shefford Cres 49 F1
Sheridan Way 48 A4
Shute End 48 D3
Simons La 48 A2
Sirius Clo 48 B3
Smiths Walk 48 B3
Snowberry Clo 48 C4
Somerset Clo 48 A3
Sorrell Clo 49 G1
South Clo 49 F4
South Dri 49 F4
Southlands Clo 49 F4
Southlands Rd 49 F4
Sparrow Clo 48 B4
Spencer Clo 48 A3
Stanley Rd 49 G3
Starling Clo 48 B4
Starmead Dri 49 F4
Station Rd 48 D3
Staverton Clo 49 H3
Sturges Rd 49 F4
Suffolk Clo 48 A3
Sundew Clo 49 G2
Swallow Way 48 B4
Swift Clo 48 B4

Tamar Way 48 B3
Tamarisk Rise 49 E2
Tangley Dri 48 D5
Tanhouse La 48 C4
Tattersall Clo 49 G4
The Green 48 A2
The Rise 48 C2
The Shires 48 A5
The Terrace 48 D3
Thorpe Clo 48 D6
Tiffany Clo 48 A3
Tigerseye Clo 48 A2
Topaz Clo 48 A2
Trafalgar Clo 48 A3
Trefoil Clo 49 G2
Trent Clo 48 B2
Tudor Clo 49 H4
Twyford Rd 48 D1

Valley Cres 48 C1
Venus Clo 48 B3
Villiers Mead 48 C3

Wallner Way 49 G4
Walnut Clo 48 B4
Walter Rd 48 A1
Ward Clo 49 F1
Warren House Rd 49 F1
Waterloo Cres 49 G4
Waterloo Rd 49 G4
Weavers Ct 49 E3
Wellington Rd 48 D3
Wescott Rd 49 F3
Westmorland Clo 48 A3
Westward Rd 48 B2
Whaley Rd 49 G1
Whitebeam Clo 48 A6
Willowside Ct 48 D2
Wiltshire Dri 49 F2
Wiltshire Rd 49 E1
Windmill Av 48 A1
Windmill Clo 48 B1
Winkfield Clo 48 D6
Wisteria Clo 48 C4
Woodland Av 48 A1
Woodlands 48 A5
Woodrows Dri 49 G3
Wooschill 48 A3
Wooschill La 48 C3
Wren Clo 48 B4

Yarnold Clo 49 H2
Yew Clo 48 B4

Zinnia Clo 48 A2